Engineering and Product Development Management

Engineering and Product Development Management is a practical guide to the components of engineering management, using a holistic approach. It will help engineers and managers understand what they must do to improve the product development process by deploying new technology and new methods of working in concurrent teams. The book takes elements from six well-known and understood bodies of knowledge and integrates them into a holistic approach: integrated product development, project management, process management, systems engineering, product data management, and organizational change management. These elements are framed within an overall enterprise-wide architecture.

The techniques discussed in this book work for both huge multinational organizations and smaller enterprises. The emphasis throughout is on practical tools that will be invaluable for engineers, managers, and consultants responsible for project and product development.

Stephen C. Armstrong is founder and president of AMGI Management Group Inc, an international operations and technology management consulting firm. He combines the business experience of a Certified Management Consultant and the technical rigor of a Professional Engineer with the practicality of a five year aeronautical engineering apprenticeship. Born in Belfast, he completed his apprenticeship in Northern Ireland with Short Brothers Aircraft, and completed a BSc Hons Mechanical Engineering degree in England at the University of Westminster – Polytechnic of Central London. Since 1981, he has worked as a design and manufacturing engineer and as a manufacturing engineering manager in North America, spending six years at KPMG as a principal management consultant specializing in advanced manufacturing systems. A consultant to some of the world's largest aerospace companies, including Lockheed Martin, de Havilland, British Aerospace, Bombardier, and Messier Dowty, Mr. Armstrong and his firm specialize in transforming business by focusing on Integrated Product Development/Collaborative Product Commerce. Their website can be found at www.amgimanagement.com. Mr. Armstrong's email address is amgi@amgimanagement.com.

Engineering and Product Development Management

The Holistic Approach

Stephen C. Armstrong

PEng, CEng, MIMechE, CMC
AMGI Management Group, Inc.

CAMBRIDGE UNIVERSITY PRESS
Cambridge, New York, Melbourne, Madrid, Cape Town, Singapore, São Paulo

Cambridge University Press
The Edinburgh Building, Cambridge CB2 2RU, UK

Published in the United States of America by Cambridge University Press, New York

www.cambridge.org
Information on this title: www.cambridge.org/9780521790697

First published 2001
This digitally printed first paperback version 2005

A catalogue record for this publication is available from the British Library

Library of Congress Cataloguing in Publication data
Armstrong, Stephen C.
  Engineering and product development management : the holistic approach /
Stephen C. Armstrong.
      p.   cm.
  ISBN 0-521-79069-7
  1. Production management.   2. Production engineering.   3. New products.   I. Title.
  TS155 .A683   2001
  658.5 – dc21   00-064218

ISBN-13 978-0-521-79069-7 hardback
ISBN-10 0-521-79069-7 hardback

ISBN-13 978-0-521-01774-9 paperback
ISBN-10 0-521-01774-2 paperback

Contents

List of Figures and Tables

Forewords

The vision of engineering management presented by Stephen Armstrong is one that is both broad in its context and deep in its coverage. He offers the engineering project manager an extensive set of management tools that, when used in total, will assure project success while improving overall project engineering effectiveness. Managers that employ this methodology will soon find this to be their indispensable desktop reference manual as they progress through the phases of product development.

The demands on the modern engineering manager are greater than they have ever been and the challenges to program success continue to grow exponentially. The rapid growth of technology has resulted in most of the products being developed by current and future companies – large and small – being inordinately complex systems of integrated technologies. This complexity is exacerbated by the complicated interdependencies among the technologies of the various product components. The availability of highly capable e-design, e-analysis, and e-prototyping tools and the growth in new methods that better integrate design and manufacturing are both wonderful benefits and potential burdens to the engineering teams using them. The move to virtual prototyping changes the planning and staffing profiles from that of the traditional project engineering organization. Added to these changes are the increasing demands for shorter and shorter engineering span times accompanied by the further expectation that engineering costs must be reduced by factors of 30 to 50 percent for businesses to remain competitive, and in some cases these reductions are expected to be recurring. These factors bring additional uncertainties and risk to an activity that has traditionally been risky.

Given this backdrop, Stephen Armstrong urges us to view the engineering management problem from a different perspective from what has been offered before. Engineering managers should adopt a total perspective of the problems that they have facing them. Even though they divide the work along the logical lines of work breakdown, they must also undertake the effort with the right tools and processes to ensure success. At the core of these processes are the ones that provide a logical and systematic definition of workflow and that provide the mechanisms to control and manage risk. Because an engineering effort is simply the maturation of information, understanding the flow of information and the management of it is critical to success. We are also cautioned that the answers to good engineering management are more than just technical or administrative. The engineering manager must recognize that his or her primary resource is people and must provide a human side to the management of engineering teams.

The managers who read this book will find the formula for successful projects. They will find useful management methods – a pattern starts to unfold and the powerful concept of an integrated technical management will form. Their approach to successful engineering management will never be the same.

Dr. Woody Sconyers, PhD
Director, Virtual Product Development
Lockheed Martin Tactical Aircraft Systems
Fort Worth, Texas

Stephen Armstrong has presented a holistic and structured approach to engineering management. It is customer focused, dealing with processes, people, communication, and their working relationships. This approach is comprehensive and offers the engineering executive an extensive and structured methodology that brings together an integrated team to enable extraordinary project success in terms of quality, cost, and schedule.

I was introduced to a holistic view of engineering management through Stephen Armstrong. This approach was first implemented at Bombardier de Havilland on the Lear 45 Wing Program with great success. Later, Bombardier Aerospace used a holistic philosophy and applied it corporation-wide. With the implementation of this management methodology, it has developed into their superb Bombardier Engineering System. It is currently being applied on new Bombardier programs such as the BD100 Continental jet.

Twenty-first-century customers are becoming significantly more sophisticated and are demanding shorter and shorter product development times, higher quality, more product performance, and lower cost. This is an ongoing challenge. The product must meet design expectations the first time. Today, customers are virtually demanding zero tolerance. To add to this challenge, new products are being developed more and more by corporate consortiums and partnerships that are faced not only with developing their components but with integrating them into the final product. Most of this is now being done with e-tools. Not only are e-tools subject to their own continuous development, but they must also be integrated with partners who are faced with a continuous training program to use these tools. The risks are increasing dramatically, and the complexity of managing all of this has become formidable indeed.

In this book, Stephen Armstrong presents a management methodology that will enable success with programs of all sizes. He takes the disciplines of integrated product development, project management, process management, systems engineering, product data management, and organizational change management and integrates them into a holistic approach for managing engineering and product development. He treats the most important constituent of a program – the people and the organizational culture. This methodology is documented in a simplified way that can be easily understood and employed by all levels of management. Knowing and understanding the information flow, workflow, and human aspect is paramount for the success of any team.

Success will come to those who read and implement the methodologies presented in this book. The material is presented with a logical flow. It provides the breadth of knowledge and the tools needed. It will lead one to the structure, organization, and effective management of a team that will make the changes required.

Carl Gerard, P Eng, MSc Eng (Cranfield)
Vice-president Engineering 1992–7 (Retired)
Bombardier Aerospace de Havilland

Preface

As the frontiers of technology advance and the work of engineers takes on an increasingly important role in our economy, companies with effective product development and engineering processes will be poised to create value for their shareholders. Those without the will to improve engineering and product development processes will be destined to lag behind.

Our university engineering programs focus on graduating technically sound engineers. Students study the disciplines of structural design or fluid mechanics. However, in both North America and Europe, little attention is paid to teaching the practice of engineering management. Engineering programs typically contain a fourth-year course on engineering economics, where students are taught the mechanics of discounted cash flows and budgets. The courses do not deal with the challenges of managing complex engineering-driven companies. With this gap in the training of engineers, it should come as no surprise when a graduate engineer practices engineering for two or three years and then leaves the profession to take an MBA. Many of these bright young engineers cut all ties to engineering. However, MBA programs are not designed to create engineering managers. The best of them teach the integration of management disciplines to teach general management; however, the worst provide the engineer with little more than a few specialized tools to apply in the area of marketing or finance. Generally speaking, the practice of engineering management is not taught in our universities. It is not a major area of research and learning, but it is vitally important to the success of today's technically driven enterprises. This problem is being addressed. Courses are being added, and enrollment is strong. The research base is lean, but certainly this book will help to fill the void.

The engineering manager at all levels has a very complex task. Just as the general manager must integrate marketing, engineering, operations, and finance, the engineering manager has an equally broad, equally complex task. Many engineering departments have specialists who have developed knowledge of a specific element of technical management. However, in today's environment, the management team must be able to look at problems from a broad, holistic perspective. To be truly successful, engineering managers must learn to integrate the concepts of a broad area of technical management disciplines. The engineering manager will need to mobilize his or her organization around this new approach. Only then will the goal of delivering new programs cheaper, faster, and with higher quality than ever before be realizable. In this book, the author takes elements from six well-known and understood bodies of knowledge and integrates them into a holistic approach for managing engineering. The disciplines of Integrated Product Development, Project Management, Process Management, Systems Engineering, Product Data Management, and Organizational Change Management are usually considered distinct, and often their implementation winds up with disastrous consequences. Never before has one integrated system been proposed to manage an engineering department from a holistic standpoint. The approach described in this book will help managers develop new products or improve existing ones faster, more cheaply, and with higher quality than ever before.

We believe that this book will provide you with the breadth of knowledge and the practical tools necessary to lead just such a change. Is this a daunting task? Perhaps, but we will address the changes required, with the same structured approach that we will learn to use to manage your new product development programs. Large problems will be broken down into manageable chunks, and suddenly they will seem very manageable indeed.

The author has been able to put a fine point on the problem after more than ten years as a consultant to large engineering organizations. In this practice, he has worked with the engineers on the CAD system improving a single workstep, all the way to the CEO in the boardroom setting a vision for an entire organization. This experience has given him a unique perspective on the problem we have just described. He knows intimately the individual management tools, but he also knows how to make them fit into a cohesive holistic plan that executives can describe but don't know enough details to implement.

The author has been involved in process management and integrated product development pretty much from its inception. In 1988–9 as a consultant for Ernst & Whinney, he facilitated the team that designed and implemented the integrated product development approach at McDonnel Aircraft in St. Louis, Missouri. McAir utilized this approach to conduct product improvements on both the Harrier and F-18 programs.

In 1991, Ernst & Whinney merged with KPMG Peat Marwick in Canada. And this led to a major business transformation assignment at Boeing de Havilland in Toronto. Bombardier Aerospace acquired de Havilland in 1992. The author received further assignments, which tended to be fundamental improvement projects to deliver step changes in organizational performance. During this time, the author left KPMG to found AMGI, the organization of which he is president today. His work at Bombardier led to the creation of the Bombardier Engineering System or BES. Building on the earlier work at McAir, the BES brought integrated cross-functional design teams to a traditional "over the wall" design engineering process. The greatest challenge on the BES was the aspect of managing organizational change within the project. The BES team fostered a common process across three countries and four cultures. Each company had the pride of its engineering heritage, bolstered by a nationalistic pride that comes from being a "national aerospace company." Today, Bombardier has applied the BES successfully on the Regional Jet RJ700, on major components of the Lear 45 business jets, and on the new Dash 8–400 regional commuter aircraft.

The author completed several assignments at the world's major military aircraft manufacturers in the period from 1996 to 2000. He assisted in the development and deployment of integrated product development to several military aircraft programs. The concept of concurrent product and process development stuck with him and has helped set the basis of the processes surrounding collaborative projects involving several partners working on a single design.

Being involved with integrated product development from its inception provides a unique perspective. The U.S. defense industry moved quickly to implement IPD, with mixed results. Typically they were trying to drive IPD separate from the other dynamics within their organization.

AMGI switched focus and began to develop a holistic approach to engineering management. Many companies will pick an initiative from one of the common management approaches. They will attempt to implement integrated product development, project management process management, systems engineering, or product data management, often with disastrous consequences because the rest of the organization actively resists the change. The holistic approach described here is unique, however, because it makes sense. Of course, new cross-functional processes are needed to support the implementation of cross-functional teams, but change of this magnitude takes vision and leadership to implement successfully. We believe that this book will provide the breadth of knowledge and the practical tools necessary to lead such a change. Top executives in most of the companies that the author has consulted have expressed their neglect of the human issues when deploying IPD or process management.

The approach documented here is a proven winner. It integrates the best thinking in the field of engineering management. Over the past ten years, we have had tremendous success putting our mark on the engineering processes of such successful engineering enterprises as Lockheed Martin Tactical Aircraft, British Aerospace Military Aircraft, Bombardier Aerospace, McDonnell Douglas, and Messier Dowty, as well as many smaller enterprises such as Ontario Store Fixtures.

This book aims to describe a straightforward model for organizing and running an engineering program and to suggest guidelines for selecting and dealing with the most important ingredient in any program, its people, and the collective organizational culture.

With the birth of e-engineering, many smaller companies are examining their product development processes. The danger is that they will fall into the trap of developing a purely electronic process. We believe that the approach we outline in this book is a prerequisite for making the move to electronic, collaborative projects. The book does not dwell on technology. Instead, it deals with people, politics, processes, and management. No technological solution will succeed if it does not consider the impact that solution will have on people. Electronic file sharing is useless if no one knows who has the authority to approve a drawing, or worse yet if the previous signatory is upset that a "team" now triggers the sign-off. These are the issues we deal with in this book. Software teams will get the electronic system up and running. Only a leader with a broad vision can make it work.

Acknowledgments

This book is based on the experience the author gained from apprenticeship in the 1970s through to the management consulting assignments performed from 1988 to 2000. Many special people have influenced, inspired, and encouraged the author to improve constantly both personally and professionally. And others in senior executive leadership positions have had the courage to risk adapting new management systems. They did this despite resistance from the established culture. Many years after an innovation is launched the original pioneers are often forgotten in the politics of change, but they are the true leaders. The following people deserve special thanks for their efforts. I am a better person for knowing them.

Technical Editing

Jim Saunders, Professional Engineer and Business Executive, for editing the book through three versions since 1997. His in-depth practical understanding of IPD and engineering management made the book possible. Jim led the original design of the Bombardier Engineering systems as an employee at de Havilland and fostered the adaptation of BES to the corporate level (1993–6).

Manuscript Typing and Layout

Jaswinder Dehal for typing endless modifications to the manuscripts from 1998 to 2000. Her precision, dedication to get the job done on time, and quality of work have been a blessing.

Jan Bowins for typing the manuscript from 1996 to 1998.

Marlene Warnysky for typing the initial manuscript from 1993 to 1996.

Consulting Assignments

Messier Dowty, Toronto, Canada, 1995–2000

Ken Laver, President, for being an executive with vision and the wisdom to adapt advanced process management methods throughout the Canadian operations of the Messier Dowty enterprise. As President of de Havilland in 1993, he sowed the original seed that led to the Bombardier Engineering System. This was later sponsored by the new President Gaston Hebert. The original sponsors of a successful change initiative are often forgotten years after it has become a way of life in a company.

Lockheed Martin Tactical Aircraft Systems, Fort Worth, Texas, 1997–2000

Dr. Woody Sconyers, Director, Virtual Product Development, and Dr. Jack Garner, Manager, Engineering Processes, for leadership in pushing the integration of process management with Virtual Product Software tool development. Dr. Sconyers provided valuable input into this book as a reviewer.

Computer Science Corporation, Cincinatti, Ohio, and Dallas, Texas, 1996–8

David Howells, Partner, for having faith in my work by introducing me to two major aerospace clients and supporting my work. In addition, Dave provided valuable input into this book as a reviewer. My work with Dave originated on the McDonnel Aircraft IPD project in 1989. And since then he has become an authority in engineering systems in the United States.

Bombardier Aerospace – de Havilland, Toronto, Canada, 1991–6

Carl Gerrard, Vice President, Engineering, for demonstrating leadership and adapting process management in engineering and product development for the first time in the company's history. This was so successful it became a corporate initiative and later an institutionalized system.

Jan McDonald, BES Coordinator, for her commitment, enthusiasm, and process excellence and for coordinating the development of BES in the early stages.

Jim Schwalm, President, for outstanding leadership in operations management and for initiating a business transformation program. His vision and hands-on approach have had an impact on Bombardier long after he was gone.

British Aerospace, Military Aircraft Division, Warton, Lancashire, UK, 1996–9

Ross Bradley, Director, Eurofighter and the OEI Transformation Program, for showing leadership in questioning the status quo and for embracing the need to address the softer human issues in process management, particularly with the Integrated Product Development initiative.

McDonnel Aircraft, St. Louis, Missouri, 1988–9

Bob Riley, Chief Program Engineer, AV8B Harrier Program, for being an inspiration when we developed the IPD approach for McAir. This was a first in the aerospace industry. The work at McDonnel Aircraft was the seed for the development and implementation of integrated product development systems at both British Aerospace Military Division and Bombardier Aerospace.

Employers

KPMG – Peat Marwick Stevenson and Kellogg (formerly Ernst & Whinney), Toronto, Canada, 1988–93

George Russel, Partner, for coaching me in the practice of management consulting and through the Certified Management Consulting (CMC) process. He is a truly outstanding mentor in the profession and I was blessed to have worked with him.

Ernst & Whinney, Cleveland, Ohio, 1988–9

Larry Michaels, Senior Manager, for coaching me in my first major aerospace consulting assignment at McDonnel Aircraft. This was truly an inspiring project and my first at applying the IPD philosophy in engineering and product development.

BBC Brown Boveri, Toronto, Canada, 1983–6

Roland Knoblauch for coaching and mentoring me in my first job in manufacturing systems and for learning the MRP II philosophy. This was a major career shift from engineering design to management.

Short Brothers Ltd., Belfast, Northern Ireland, 1972–7

Liam Begley, Senior Tool Development Engineer, for mentoring me through my apprenticeship when I moved from the shop floor into the manufacturing engineering technical department.

A. B. Treacher, Chief Planning Engineer, and Ernie Crone, Planning Engineer, for encouraging me to continue my studies and inspiring me to study engineering at the university level. I am indebted.

Publishers

Dr. Philip Meyler, Senior Commissioning Editor, Cambridge University Press, Cambridge, England, for helping me through the initial preparation phase of the publishing process.

Milicent Trealor, Editor, Society of Manufacturing Engineers, for encouraging me to approach Cambridge University Press and for coaching me through the process.

Family

Various family members in both Northern Ireland and the United States for providing me with a solid foundation in which to carry out my work and for showing faith in me throughout the years. Many are deceased but will never be forgotten.

Layout of Book at a Glance

Part 1 – Learn the underlying body of knowledge

Part 2 – Learn the tools and techniques of engineering management

Part 3 – Learn how to make use of this knowledge in your organization and overcome resistance

PART 1 – UNDERSTANDING ENGINEERING PROCESS MANAGEMENT

1 – THE HOLISTIC APPROACH TO MANAGING ENGINEERING OPERATIONS
- Bodies of Knowledge
- The Holistic Approach
- IPD Philosophy
- The Integrated Enterprise Framework

2 – AN OVERVIEW OF ENGINEERING PROCESS MANAGEMENT
- Process Decomposition
- Customer Deliverables
- Maturity Gates
- Process Maturity

3 – ORGANIZATION OF ENGINEERING TASKS
- Tracking System
- Integrated Master Plan
- Integrated Master Schedule
- Work Plan Templates

PART 2 – APPLYING ENGINEERING PROCESSES TO PROGRAM MANAGEMENT

4 – ROLES AND RESPONSIBILITIES
- Customer
- Partner
- Sponsor
- Functions
- Project Manager
- IPTs

5 – APPROACH TO PROGRAM AND PROJECT MANAGEMENT
- Program Office
- Success Factors
- Program Modules
- Framework
- Soft Side
- Hard Side

6 – AN INTEGRATED TEAM MEMBER'S GUIDE TO PERFORMING A TASK
- IPT Member Responsibilities
- Work Products
- Performing Tasks

7 – PROGRAM STRUCTURING AND PLANNING
- Approach/Benefits
- Structuring
- Tailoring
- Integrated Master Plan
- Review Plan

8 – RISK ASSESSMENT
- Performing an Assessment
- Strategies for Risk Management

9 – PROGRAM INITIATION AND EXECUTION
- Setting Goals
- Training
- Team Management
- Environment
- Work Management
- Re-planning

10 – PROGRAM REVIEWS
- Q & A Factors
- Tailoring
- Measuring Quality
- Q & A in IPD
- Review Process
- Review Board

11 – ENGINEERING CHANGE AND PDM
- Engineering Change Concepts
- Engineering Change Process
- PDM Concepts/Benefits/Justification

PART 3 – DEPLOYING ENGINEERING PROCESS MANAGEMENT

12 – ORGANIZING FOR DEPLOYMENT
- Program Organization
- Charter & Objectives
- External Consultant
- Budgeting
- IPD Plan
- Sector v. Corporate

13 – OVERCOMING RESISTANCE TO CHANGE
- Political Management
- Adapting Teams
- Obstacles to Effectiveness
- Overcoming Resistance to Change Approach
- Leadership
- Behaviors

14 – IMPLEMENTING IPD – LESSONS LEARNED CASE STUDY
- Leadership
- IPT Setup
- Decision Making
- Roles & Responsibilities
- Communications
- Team Skills & Training

Part 1

Understanding Engineering Process Management

1 The Holistic Approach to Managing Engineering Operations

This book is intended to help you implement a more rigourous approach to the practice of engineering management. In our consulting work, we have seen many attempts to improve this management process. Mostly we have been called in when previous initiatives have failed. We believe that these initiatives have failed for one simple reason. They were narrow, one-dimensional solutions to problems that had many facets. Additionally, the people who offered the consulting services quite often had very little practical background in engineering. In this book, we will look at the engineering process from a holistic approach.

Typically, we see the scenario play out as follows: a firm finds that its development projects are taking too long or costing too much money to complete. A senior manager in engineering has read a book, attended a course, or acquired considerable experience in one particular approach and recommends that the organization simply implement this new way and the problem will be solved. In the best case, a year later the performance in the one area at which the solution was aimed has improved, but there is no broad-based financial measure of improvement. In the worst case, the organization is in upheaval with pockets of resistance firmly entrenched against the change.

The reason for this failure is the application of improved methods in isolation from one another. These solutions typically originate from one of six bodies of knowledge. A *body of knowledge* is an inclusive term that describes the sum of knowledge within a profession or management practice. A body of knowledge includes knowledge of proven, traditional practices, which are widely applied, as well as knowledge of innovative and advanced practices, which may have seen more limited use. The six bodies of knowledge that we examine are fundamentally sound and broadly accepted in modern management theory. In this book, we will look at how to integrate them into a cohesive approach to making quantum improvements in your product definition process.

In this first chapter, we will examine the six bodies of knowledge that we will integrate to form our approach. We will look at the motivation that led us to take this tack. That discussion will lead us to examine the benefits of the holistic approach. Then we will look at some of the critical success factors that will make the difference as you move to implement it in your organization. Finally, we

3

will introduce what we call the Integrated Enterprise Architecture as a model for managing broad based change programs.

1.1 SEPARATE BODIES OF KNOWLEDGE

Top managers have two perspectives that separate them from the middle managers. First, the language of middle managers is "things," whereas the language of top managers is dollars. Certainly, top managers have a way of looking at issues from a financial perspective. However, top managers also have a wide breadth of perspective. Managers will usually rise to the top on their ability to synthesize the input from many different perspectives. They lead their organization on a path that will optimize many, many different variables. This book was written to help give you that breadth. It will help you look at improvements to your product development process from the same perspective as the most successful executive. Then it will help you set a course of action that optimizes all the variables and constraints. It will be particularly useful for engineers and scientists who aspire to executive management. The traditional education and training of engineers is based on the applied sciences. Scientific training has tended to create people with black-and-white or right-and-wrong thinking. This thinking works well when dealing with equations and the design of things, but it falls apart when dealing with people. Unlike things, people exhibit unpredictable behavior. The interaction of politics and emotions is the greatest challenge management faces. And the transition from engineer/scientist to manager is a major challenge. This transition can work only if the executive takes a holistic approach at applying what are traditionally separate disciplines. We will develop a practical approach that can be implemented readily.

This approach, while broad, draws its strength from a grounding in six long-trusted management bodies of knowledge (see Figure 1-1). In the past middle managers have tended to treat them as separate entities possible because they each have their own professional bodies that promote them. Middle managers have tended to lock onto one body of knowledge to call their own and then to develop an expertise in that area. For example, they may have led a crusade to implement business process reengineering, concurrent engineering, systems engineering, or project management. Our approach in this book, however, is to integrate these previously separate disciplines. We believe that this approach has not been documented in management literature.

Let's first examine the foundation, the six bodies of knowledge upon which we base our approach.

Integrated product development

Integrated product development (IPD) was the brainchild of the U.S. defense industry in the late 1980s. In the IPD approach, engineers form multidisciplinary

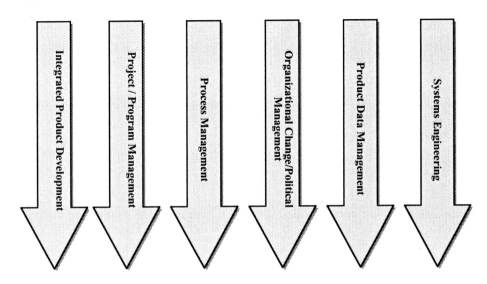

Figure 1-1: The six management bodies of knowledge

teams with the mandate to deliver a product that optimizes all the individual constraints placed upon it. These constraints could include budgets, materials, speed, strength, and functionality, among others. The list could go on indefinitely. However, for the purposes of this book, and to promote a common understanding of IPD, we have adapted the definition provided in the U.S. Department of Defense *Guide to Integrated Product and Process Development.*

Integrated product development is "[a] management process that integrates all activities from product concept through production/field support, using a multifunctional team, to simultaneously optimize the product and its manufacturing and sustainment processes to meet cost and performance objectives."

IPD replaces the serial or "over-the-wall" approach in which a designer completes a design and throws it over the wall to someone in an analytical discipline who, in turn, throws it back saying something like, "It needs to be stronger." As ridiculous as it seems, most design departments still operate this way. IPD utilizes integrated product teams (IPTs) to allow all affected areas to provide their perspective up front. The IPD approach allows the specialists to collaborate on the design and collectively present a finished product to a manufacturing team that already knows exactly what it will be receiving.

Concurrent engineering is at the heart of IPD. The IPTs are multifunctional teams organized around the major components of the product (or product structure). These teams have responsibility for developing a fully functional, producible, highly reliable product that meets or exceeds customer requirements, program budget, and schedule constraints. The teams have authority over all technical, cost, schedule, and quality aspects of their products. They are accountable to the respective program manager. All IPT team members represent the expertise of their functional discipline and are able to provide input to team decisions based on the processes and procedures developed and maintained by their functions. The multifunctional nature of the IPTs ensures early consideration of all issues relevant to cost, schedule, and performance of the product throughout its life cycle. The IPTs should be formed early in the proposal phase of a program and should retain their focus throughout the life cycle of the product. This early involvement of all functional disciplines that are vested in the development and support of a product is the defining difference between serial and concurrent approaches to managing a program.

Colocation is a key enabler to the IPD process. Through the colocation of core members of the IPT, integration team, program manager, and appropriate business support resources, the free exchange and timely flow of information among program team members is spontaneous. Brain storming sessions are performed on an as needed basis.

Although this early involvement of all appropriate functional disciplines increases initial program costs when compared to the serial approach, product changes downstream are either avoided or greatly reduced, resulting in both cost and schedule savings.

Project/program management

The work of an enterprise generally involves either operations or projects. They are similar in that they are performed by people, constrained by limited resources, and then planned, executed, and controlled. Operations and projects differ in that operations are ongoing and repetitive, whereas projects are temporary and unique. The project management body of knowledge is documented in the Project Management Institute's *Guide to the Project Management Body of Knowledge*. They define a *project* as "a temporary endeavor undertaken to create a unique product or service. Temporary means that every project has a definite beginning and definite end. Unique means that the product or service is different in some distinguishing way from all other similar products or services."

The Project Management Institute defines *project management* as . . . "the application of knowledge, skills, tools and techniques to project activities in order to meet or exceed stakeholder needs and expectations from a project. Meeting stakeholder needs and expectations invariably involves balancing competing demands among:

- ○ Scope, time, cost and quality
- ○ Stakeholders with differing needs and expectations
- ○ Identified requirements (needs) and unidentified requirements (expectations)."

Project management involves planning the project, executing the plan, monitoring the execution, and wrapping up the project when complete. Good project managers would die for their projects; great project managers inspire a team that would die for their leader.

Process management

Michael Hammer's book *Re-Engineering the Corporation* launched business process reengineering in the early 1990s. The fact that you do not hear the term as frequently today will lead some to believe it was another management fad. In fact, it is quite the opposite. The e-business boom we are seeing today is simply a massive reengineering project. When we introduce the integrated enterprise architecture later in the chapter, you will see that the impact e-business is having on the movement of data is not possible without fundamental process change. An enterprise runs on a series of processes. Process reengineering involves examining in detail how the organization performs the tasks that create value. Using process thinking, we stop thinking of the enterprise as a collection of functions and begin to see it as a collection of integrated tasks that make up processes designed to create and deliver value to the customer.

These processes run across the organization moving information or physical materials from function to function (or department to department). In process reengineering, we examine these flows, document them, measure them, cut steps, add new steps, and ask the fundamental question, "Should this process exist at all?" Michael Hammer's landmark *Harvard Business Review* article in July 1990 was entitled "Reengineering Work: Don't Automate, Obliterate."

Organizational change/political management

Not so many years ago managers had time to worry about the next wave of change that they potentially would have to ride. As a manager in the new millennium, that wave of change has become perpetual white water. In managing organizational change, we utilize organizational behavior techniques to understand and systematically overcome resistance to change. Engineers in management positions often give short shrift to this "wooly side" of management. They would prefer to deal with laws of physics or physical properties of materials. The body of knowledge is as diverse as Maslow's hierarchy of needs, psychology techniques, sociopolitical management, and military strategy. On its own, this discipline is merely mental gymnastics. An improvement initiative that ignores the people aspects will certainly fall from the balance beam. The book *The Change Masters* written in 1985 by Rosabeth Moss Kanter was one of the first texts to address this body of knowledge. Its central theme focused on achieving an American corporate renaissance by stimulating more innovation, enterprise, and initiative from the people.

Product data management

Product data management (PDM) is a class of software designed to provide easy access to the mountains of data that engineers create. PDM systems provide a central repository for the design data so that authorized people can access the current design documentation with the confidence that they are using the appropriate version. PDM systems help the organization to define relationships between data elements. Any authorized person can easily locate a given component and then access all related information such as the CAD drawing, the stress analysis, the test results, and the modifications to the component that are in work, approved, or in production. Advanced PDM systems also provide the functionality to help manage the product development process. They can enforce a standard routing, known as a *workflow*, and can help project managers monitor activities through the workflow, thereby offering rudimentary project management functionality.

Systems engineering

Systems engineering is the practice of coordinating and executing development activities for designing and building systems – large or small, simple or complex. Systems engineering work begins with the needs of stakeholders, users, and operators, and transforms these needs into a responsive, operational system design and architecture. The deliverable must conform to the demands of the marketplace as well as to the initial set of functional and nonfunctional requirements.

The International Council on Systems Engineering (INCOSE) defines *systems engineering* as "an interdisciplinary approach and means to enable the realization of successful systems." More succinctly put, systems engineering is an organized approach to problem solving by an experienced engineer with a broad, systemwide overview toward solving the problem, weighing options, and evaluating risks and constraints.

Systems engineering is responsible for integrating all the technical backgrounds, subject matter experts, and specialist groups in a development effort. It starts with defining customer needs and required functionality early in the life cycle and managing requirements and proceeds into design synthesis and system verification and validation.

A major element of system engineering is risk management. Project management specialists will argue that risk management is just a chapter of the program management body of knowledge. Indeed, program directors are responsible for assessing the risks their projects face, assessing the potential impact, and taking the steps necessary to mitigate those risks. Risk, however, can have as large an impact on cost and schedule as it has on the original product requirements. We have all heard of projects that take twice as long to complete and cost twice as much as the baseline plan. On complex programs, managing risk is a major process. From a holistic approach, risk management involves identifying the appropriate senior people and gaining their commitment to assist in managing the technical and business risks on the program. Effective risk management involves controlling elements of timing, managing the political environment in the organization, and being sensitive to the needs of the team, the partners, the customer, and executive management.

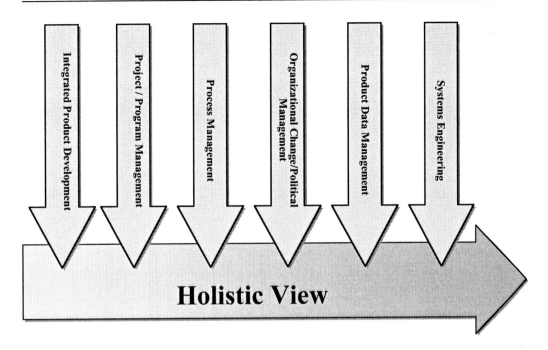

Figure 1-2: The holistic view of the bodies of knowledge

1.2   THE HOLISTIC APPROACH

In the holistic approach shown in Figure 1-2, we take the best that these discrete bodies of knowledge have to offer. We rigorously practice what they preach, but we take a step back to consider the impact our actions in one area will have on another. We will redefine the engineering process, but we will do it in a way that makes project management simpler and more effective as well as one that allows for the implementation of product data management, and so on. Throughout the book, we will identify key points where you should stop driving down one path and examine how to integrate concepts from the other bodies of knowledge to create an entire route map.

1.3   THE MOTIVATION FOR TAKING
THE HOLISTIC APPROACH

Programs or projects, regardless of their scope, size, or complexity, must be acquired and executed in a similar manner by performing a basic set of activities. A methodology is the organization of these activities into a standard framework, one that is used to guide the planning, execution, and management of programs. This methodology allows us to manage programs with repeatable results by approaching them with scientific methods and to make sure that these methods are aligned with the human aspirations and political climate.

IPD objectives

Companies that engineer products stake their future on their ability to continue to play a significant role in the development of new products. As such, we must address a basic issue faced by all engineering organizations, namely to reduce cost and overall product development cycle time and improve quality. Some typical goals follow:

- Meet program cost targets
- Reduce total program costs
- Meet program milestones
- Reduce time to market
- Reduce engineering change following design freeze

As a management tool, this book is intended to help the management team answer the following questions:

- What work needs to be completed?
- What deliverables will be produced during the process?
- What skills are necessary to complete the work?
- How long will the work take?
- How much will it cost?
- How does each piece of work relate to other pieces of work?
- When are program reviews appropriate?
- What resistance must we overcome to achieve our goals?
- How do we organize for deployment an enterprise-wide architecture?

We aim to provide a standard framework for executing engineering projects, thereby allowing your company to concentrate its efforts on applying skills (e.g., stress analysis, mechanical design, and process design) so as not to be defocused recreating the approach for each program.

1.4 BENEFITS OF THE IPD APPROACH

The main benefit of the IPD approach is that it provides an effective mechanism for planning and managing an engineering program. This benefit can be broken down into five more specific benefits:

- To help ensure high-quality product and process definition
- To improve project management and business process definitions
- To capitalize on experience gained from other projects
- To establish consistency across projects and divisions
- To provide a framework for training

1.4.1   Ensuring high-quality product and process definition

A consistent approach helps to ensure high-quality product definition through efforts involving both people and technology. The following discussion introduces these concepts.

Effective communication

High-quality product definition requires effective communication processes among project team members and between the team partners and the customer. One of the greatest benefits of a well-documented approach is that all involved in the development process speak the same language. The team members and the customer agree on what work is involved in each phase of the project. The process promotes effective communication by building regular reviews into the project. These reviews enable all key players to monitor the program's progress and to identify and correct flaws early in the development process.

Complete documentation

Complete project and system documentation is another key to ensure quality product definition. Documentation is produced throughout the course of the project. The documentation is cumulative in that deliverables resulting from one task are typically used as input for subsequent tasks. At the project's end, these deliverables become the final design documentation.

1.4.2   Improving project management

In this book, we will outline techniques for high-level and detailed project planning as well as provide additional information on risk assessment, estimating, and project execution and control.

Project structuring and planning

The starting point for effective planning is the business process framework. It provides a way to break down the design activities to help identify deliverables (called a work breakdown structure or WBS), plan major segments of work, and determine the program review points necessary for a program. At a detailed level, the process flows provide a foundation for work-planning templates and for the project WBS. These templates contain lists of tasks and work steps that represent the work typically involved in the development of an engineered product.

In addition, we will develop a template specifically for defining the project management tasks associated with a project such as the steps for planning the project, setting up the program review process, establishing program-specific standards, initiating the project, and monitoring and controlling it. By separating these project management tasks from the development tasks, we focus the project manager/engineer on the activities necessary to successfully manage a project.

Project estimating

Quite simply, the repeatable results that a consistent process can deliver lead to higher quality cost estimates. Better cost estimates lead to better financial management and profitability.

Project execution

A consistent process assists in project execution by clearly outlining what is expected from a particular deliverable. Process documentation provides a comprehensive reference library, with descriptions of all tasks, to help team members understand the work involved.

Project control

To enhance project control, the process borrows proven techniques from other disciplines. From industrial engineering, for example, we apply the principles of short-interval scheduling and tangible deliverables. From financial disciplines, we apply formal reviews to compare actual times and costs to those budgeted. This enables tighter control over project progress and provides valuable feedback to improve the accuracy of future project estimates.

1.4.3   Capitalizing on experience

To provide the best direction possible, we need to capitalize on experience gained from other development projects. The process framework and all the associated tasks and deliverables are based on years of engineering design experience and many previous projects. As technology advances and that experience in using the techniques grows, we should update it to make sure it reflects the latest practices of your organization.

1.4.4   Establishing consistency

Consistency, both across projects and between different aspects of the same project, is another objective. We outline a common work breakdown structure, to help ensure that all program and functional directors consider the same set of tasks in planning a program.

All project teams will use a consistent approach and common terminology. Common terminology facilitates the rapid transfer of personnel among programs at your company as well as partners and suppliers. As a result, the individuals with the required specialized competencies can be brought onto the project team and up to speed quickly.

1.4.5   Providing a training framework

Another objective of our work is to provide a framework for training. Because the tasks cover many aspects of engineering design and program management, they provide a natural source of material for your training curriculum. As the embodiment of your engineering process, it provides a firm foundation for training new staff. It also gives engineering professionals a common set of standards and procedures upon which to build.

1.5   OVERVIEW OF THE IPD PHILOSOPHY

All six of the bodies of knowledge we introduced are important, but Integrated Product Development is at the core of this holistic approach. It represents a fundamental change from the way engineering projects/programs have been organized and run in the past. It provides a flexible approach to simultaneous product and manufacturing process development.

At the core of IPD implementation are integrated product teams that are organized to accomplish tasks required by the project. They work together to consider all life cycle issues from initial product conception to final product delivery. The IPT environment provides a tremendous opportunity to tap into the innovative talents of each team member to improve not only the product but also the fundamental way in which it is created.

IPD is based upon commonsense decision making. It involves bringing the right people together at the right time to make the right decisions. It mandates that all IPTs use quality, cost, schedule, and technical considerations as a four-point check during each step of the design process. If a program is operating within the IPD philosophy, each team member will be able to identify his or her product responsibility, decision authority, and responsibilities on the program. While bringing additional manpower (and cost) to the beginning of the development process, IPD methodologies have been proven to "scrap paperwork . . . not product." IPD does this by shifting effort from the back end of the product development process, where changes are costly, to the front end where changes don't have an impact on production.

Within a program, IPD increases focus on and ownership of processes and products, improves horizontal communications, clarifies the interfaces between participating functions, enhances teaming through rapid, open communications, and establishes clear lines of authority and responsibility.

An effective IPD methodology focuses everyone's efforts on heightened customer satisfaction by systematic concentration of effort on the product. The way in which an enterprise organizes, plans, costs, monitors, designs, manufactures, supports, and retires a product is ultimately affected. This is done through the use

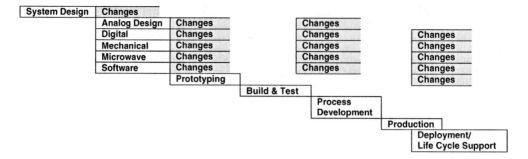

Figure 1-3: Serial approach to product and process development

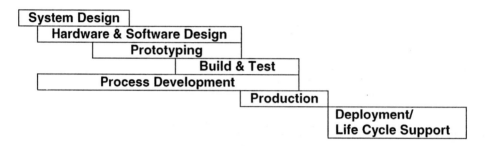

Figure 1-4: Integrated product development approach

of plans and schedules that are structured in the same way as the product itself. For example, there is a work plan and schedule for the development of the door assembly rather than a stress engineering functional plan.

1.5.1 Comparison of IPD and traditional approaches

To better understand integrated product development, we need to examine it in contrast with the alternative – the traditional serial approach. Figures 1-3 and 1-4 compare the serial approach to integrated product and process development with the concurrent approach. The important element to note when comparing Figure 1-3 to Figure 1-4 is the dramatic reduction in time because processes are performed simultaneously.

1.5.2 The serial approach to product and process development

In the serial approach, the customer provides top-level requirements during the concept definition stage. Systems engineers refine these requirements before passing them on to hardware and software engineers (see Figure 1-3). Hardware and software engineers create design concepts and document them in the form of "engineering drawings." During the product design phase, these drawings are refined and then passed along to specialty engineering organizations, purchasing, manufacturing, quality, and other support functions. These groups then perform

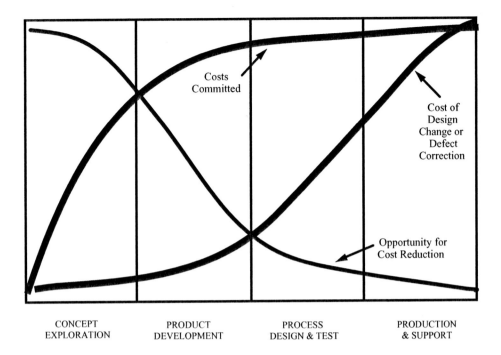

CONCEPT
EXPLORATION

PRODUCT
DEVELOPMENT

PROCESS
DESIGN & TEST

PRODUCTION
& SUPPORT

Figure 1-5: Cost impact

their respective analyses, order materials, select vendors, design and fabricate tools and test fixtures, and prepare logistics plans. At each of these hand-offs, if the recipient did not contribute to the decisions that led to the product he or she is presented, there is a probability that changes will be needed. These changes can extend the product delivery cycle, increase development costs, increase overall product costs, or result in compromises adversely affecting quality.

In product development, between 60 and 80 percent of the overall product costs are committed between the concept and preliminary design phases of the program. However, in the serial approach, only a small cumulative expenditure of funding, 5 to 10 percent, is committed during this same period (see Figure 1-4). This relatively small expenditure during these stages reflects the limited involvement of functional disciplines outside of the engineering. Because of traditional schedule considerations, the design usually gets locked in after the concept definition stage, without adequate coordination or consideration of inputs from other affected functions. This lack of early design coordination forces disciplines outside of engineering to come up with manufacturing, inspection, and support processes that may not be optimal. Often the impact is an order-of-magnitude greater when resolution of a problem is delayed to the next stage in the process (see Figures 1-5 and 1-6). The factor-of-ten rule states: It costs roughly ten times as much to fix a problem if it is delayed to the next process step. In Figure 1-5, notice how committing costs and higher effort at the concept phase dramatically reduces costs downstream at the production and support phase.

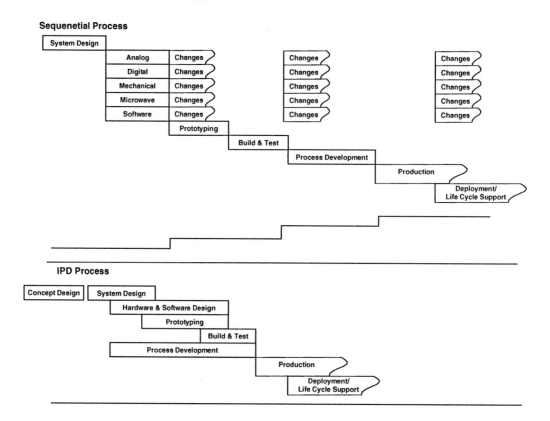

Figure 1-6: Cost impact of change

1.5.3 Team comparisons

Even through an IPT is similar to other work-related teams in many respects, it also differs from them significantly in several ways. Table 1-1 compares an IPT, a producibility team, and a tiger team, teams with which most employees are familiar.

Perhaps the most significant differences between the IPT and work-related teams are in the areas of responsibility, authority, and duration. The differences point to a need for changes in the approach to team decision making, and to the interaction among the functional organizations.

1.5.4 Empowering the integrated product team

To empower a person or team is to grant the person or team the power or authority to perform a certain function or task. An IPT is empowered to make decisions concerning specific aspects of its product within defined requirements. Empowerment of the team does not mean that management relinquishes its authority. Rather, management shares authority with the team and maintains the prerogative of periodic review.

Table 1-1: Comparison of IPT and other teams

Team parameter	IPT	Producibility team	Tiger team
Objective	Product delivery	Cost-effective production	Problem solving and proposals
Orientation	Design for life cycle (proactive)	Ensure that design is producible (reactive)	Solve the problem (reactive)
Responsibility	Deliver product to realistic specifications at low cost in minimum time	Recommend changes to save manufacturing and inspection costs	Recommend and implement solutions
Authority	At team level	Functional managers and designers	Team leader and functional managers
Leadership function	Facilitate, coordinate, spokesperson	Coordinate	Coordinate, direct
Duration	Concept design from proposal to disposal	Short term, starting after detailed design	Short term for duration of problem
Structure	Multifunctional teams addressing product systems	Multifunctional teams critiquing key designs	Multifunctional teams solving a specific problem

The success of an IPD approach to product design can be attributed to IPT empowerment and synergy. Synergy results from assembling a team capable of working together to ensure that all product life cycle requirements are addressed during upfront design – an effect that could not be accomplished by team members working only as individuals. Periodic program reviews ensure that an IPT is addressing and meeting their individual product accomplishment and performance criteria, as well as program requirements, plans, and cost targets.

The main difference between an IPT and other types of teams is that an IPT is given overall responsibility for and authority over a specific product and its associated processes, throughout its life cycle. The IPT is given comprehensive responsibility for the development of a system or subsystem, consistent with product design requirements. This includes responsibility for managing team resources in implementing the development program. The IPT's continuity, responsibility, and authority empower it to do its job of delivering a reliable, maintainable, high-quality product on time and at the lowest cost.

1.6 CRITICAL SUCCESS FACTORS IN IMPLEMENTATION

To implement a new engineering process effectively, it is important to understand the interrelated characteristics. The following characteristics are a compilation of best practices at various engineering companies. They represent the core principles and themes, which embody a truly revolutionary environment.

○ *Customer focus*

The primary objective of the integrated product teams (IPTs) is to satisfy the customer's need for better and faster at a lower cost. These needs include product specs, delivery dates, and quality levels. The customer's needs should determine the nature of the product and its associated processes.

○ *Concurrent development of products and processes*

Processes should be developed concurrently with the products they support. It is critical that the processes used to train people and to manage, develop, manufacture, verify, test, deploy, operate, support, and eventually dispose of the product be considered during product design and development. Product and process design and performance should be kept in balance. Development decisions must be directed and viewed with respect to the impact on the product life cycle.

○ *Early and continuous life cycle planning*

Planning for a product and its processes should begin early in the underlying research for the product and extend throughout the product's life cycle. Early life cycle planning, which includes examining the roles of customers, functional organizations, and suppliers, lays a solid foundation for the various phases of a product and its processes. Key program events should be defined so that the resources can be applied and the impact of resource constraints can be better managed. The program milestones should be documented in the form of an integrated master plan (IMP), containing associated criteria for successful route map, dependencies, IPTs responsible, and supporting Statements of Work (SOWs).

○ *Maximize flexibility for using a subcontractor and partner*

Requests for Proposals (RFPs), proposals, and contracts should provide maximum flexibility to deploy IPTs, contractor/partner unique processes, and commercial specifications, standards, and practices.

○ *Encourage robust design and improved process capability*

Advanced design and manufacturing techniques are used to promote quality through design, products with little sensitivity to variations in the manufacturing process (robust design), and focus on process capability and continuous process improvement. To reduce process variability, such tools as "six-sigma" and lean/agile manufacturing concepts are used.

○ *Event-driven scheduling*

A scheduling framework that relates program events to defined criteria should be established. This is the basis of the IMP. This event-driven scheduling reduces risk by ensuring that product and process maturity is demonstrated prior to beginning the next activities and should be documented in the form of an integrated master schedule (IMS). Every event that appears in the program IMP must be included in the program IMS. The IMS then serves as the master schedule for all detailed tasks of the IPTs.

○ *Multidisciplinary teamwork*

Multidisciplinary teamwork is essential to the integrated and concurrent development of a product and its processes. Team decisions should be based on the combined input of the entire team (engineering, manufacturing, test, financial management, customers and suppliers, etc.). All team members need to understand their own roles and to support the roles and constraints of the other members. The manpower planning for the program should be driven by the program IMP and IMS. The planned skill mix must be supported by the functional area managers.

○ *Empowerment*

Decision making, authority, responsibility, and resources to manage their product should be driven to the lowest possible level commensurate with risk. Decisions include establishing budgets, schedules, and resource and facility planning. The team accepts responsibility and is accountable for the results of their efforts. This may be documented in the form of team agreements, a recognized best practice.

○ *Seamless management tools*

A framework that relates products and processes at all levels to demonstrate dependencies and interrelationships should be established. A single management system that relates requirements planning, resource allocation, execution and program tracking over the product's life cycle should be established. This integrated approach helps ensure that teams have all available information thereby helping the team to make decisions. These tools help ensure that all decisions are optimized toward the customer's needs and the product's life cycle goals.

○ *Proactive identification and management of risk*

Critical cost, schedule, and technical parameters related to system characteristics should be identified from risk analyses and user requirements. Technical and business performance measurement plans, with appropriate metrics, should be developed, collected, and reviewed to track achievement of technical and business parameters.

○ *Management commitment*

The change initiative begins with management from the manufacturer, partners, and customer committing to the new process. Management in this sense is not limited to program management but includes the functional and support organization management as well. To be successful, a willingness to embrace this philosophy must run throughout all organizations, strategic partners, sites, and functional organizations. The absence of management commitment is the major reason change initiatives fail.

○ *Communication*

Efficient and timely communications, data flow, and collaboration to make decisions and perform tasks are essential. Communications must be open within teams as well as between teams. Improved communications will always mitigate risk.

○ *Continuous process improvement*
The formulation and tracking of appropriate IPD process metrics are essential to appraising and improving product development policies, standards, and processes.

○ *Integrated product team leaders*
Careful selection and training of IPT leaders, members, and facilitators is a key starting point. It is very important to provide IPT leaders with leadership training, meetings management, consensus building, and the like.

○ *IPT environment*
The success of the new engineering process depends on the environment in which the IPTs must function. In a healthy environment, managers assume the roles of facilitator and coach of their team members. Additionally, they must assume ownership of the functional processes that support the program. Managers will be called upon to sponsor and champion teams, acting to remove barriers to success and empowering teams to get the job done within specific boundaries.

1.7   THE INTEGRATED ENTERPRISE FRAMEWORK

The process framework of an enterprise is only one element of its infrastructure. Top management at any enterprise should develop an integrated enterprise framework. The integrated enterprise framework, as shown in Figure 1-7, is a way to help people keep the principal structures or domains of change of the business in balance, as they change the way work is performed. It ensures that implementation plans cover all aspects (e.g., training, attitudes, technology, location, and organization). The principal domains of change of an enterprise are

○ People/culture/philosophy
○ Business process structure
○ Physical structure
○ Computer/technology structure
○ Information/knowledge structure

For example, e-business initiatives focus primarily on redefining the business process and information structures of the enterprise. The Internet allows different companies to collaborate using common data to improve processes, improve service, and reduce costs dramatically. At this point, it is also important to understand that we will not be looking at strategies for implementing e-business solutions in this book. The framework we have presented transcends technology. We deal with the management process. The operational processes of how to conduct a

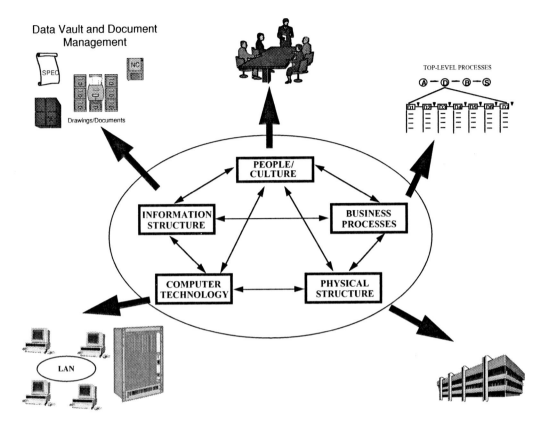

Figure 1-7: Integrated enterprise framework

stress analysis or the right data protocols for sending design documentation to a partner is the subject of another text. There is a great deal of hype on e-business or e-engineering. No technology ever has or ever will improve business performance without an efficient process that fosters human innovation within a politically aligned organization.

We will deal with strategy. The strategy for any enterprise should contain two components. First, there should be an external strategy. There are many good texts on business strategy. For example, Michael Porter's *Competitive Strategy* is excellent. In this text, however, we will focus on developing an internal operations strategy. Implementing a new engineering management process then is a major element of the operations strategy. As such, any implementation should consider the degree of fit or misfit between the new processes, the concurrent approach, and the IPTs and domains of change described earlier. For example, implementing IPTs will be much more complicated in an organization that has a history of autocracy. Similarly, concurrent engineering will create substantial amounts of corporate data. Therefore, the support of the information technology organization will be crucial for supplying the necessary horsepower.

Before an enterprise can design future state processes, management should develop a process framework that covers the entire enterprise. The process framework

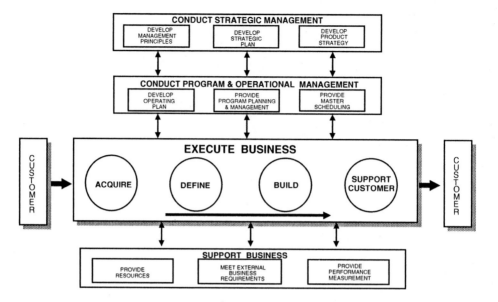

Figure 1-8: Top-Level business processes – example

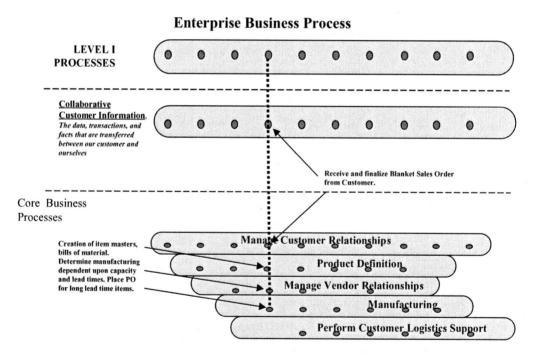

Figure 1-9: Top-level process framework – example (Courtesy Ontario Store Fixtures, Inc.)

identifies and documents process boundaries, inputs, executions, and outputs. This architecture is used as a common framework for improvement efforts. The top-level process model shown in Figure 1-8 encompasses all work at an enterprise including

○ Conduct strategic management
○ Conduct program and operational management
○ Execute business
○ Support business

In Figure 1-8 the execute business process is where we focus the majority of our attention in further developing processes that are efficient and effective. The execute process is often referred to as the core process. The core processes (acquire, define, build, and support customer) need to be broken down, or decomposed, at least two levels deeper. In some cases, they should be further decomposed to the task and workstep level. In practical terms, you cannot improve a process unless you analyze the level where work gets done. Figure 1-9 identifies a top-level process framework that incorporates the external customer processes. In this example, we demonstrate how internal processes must be aligned and coordinated with customers and suppliers in the "supply chain."

2 An Overview of Engineering Process Management

In Chapter 1, we introduced the six bodies of knowledge that are the foundation of good engineering management. This chapter outlines the key elements and critical steps necessary to implement process management concepts. We will use a process framework to structure the product definition engineering (define) process. Tailoring these processes should be based on customer, program, and business needs. However, certain fundamentals such as multifunctional teams, timely communication, integrated master plans, integrated master schedules, team colocation, and empowerment, should be applied on all of your company's programs.

This chapter looks at how to structure the Define process to help achieve business objectives. To this end, we discuss the work breakdown structure, the process framework, the structure of customer deliverables, the use of product maturity gates to control design quality, and the organization of engineering documentation.

2.1 ENGINEERING PROCESS FRAMEWORK

The approach to engineering design is based on aligning activities to the process framework. Years of experience have shown that every engineering project follows a similar path from initiation to completion. We know that no two projects or programs follow exactly the same path; however, defining this general process framework provides the foundation for engineering process management, upon which the detailed components have been developed.

Table 2-1 (see Appendix B) shows a process framework with its objectives and major deliverables (ex: Design Documentation) for each of the phases, and Table 2-2 (see Appendix B) identifies the descriptions and deliverables for each of the subphases in the process framework.

In this book, we will focus on the Define process. We will break the Define process down first into phases. The first level, Tier 1, is illustrated in the example in Figure 2-1. We will use this model throughout the book to illustrate practical techniques. This example identifies the seven phases and twenty-six subphases. Figures 2-2 and 2-3 illustrate an alternative Tier 1 process structures.

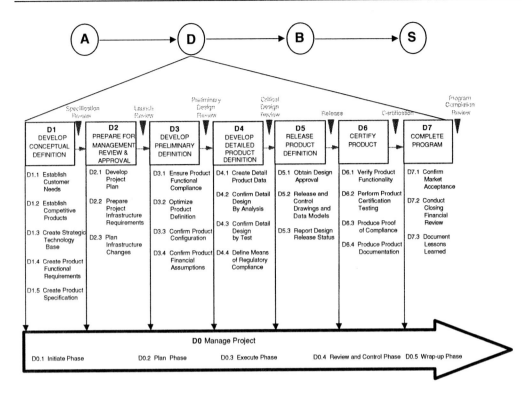

Figure 2-1: Process framework example 1 – Tier 1 processes (Courtesy Bombardier Aerospace – de Havilland Division)

In Figure 2-2, the phases define the product life cycle. In this example, the framework defines product maturity in seven discrete phases (A to G). The framework decomposes into twelve overlapping processes illustrated in boxes that can be applied to all projects. Figure 2-2 illustrates processes that overlap and cut across phases. This contrasts with Figure 2-1 where processes are structured phase by phase. The phases are controlled by "gates," which control the progression of the project through the life cycle. An understanding of the phases will enable the project/product to be positioned (e.g., phase C or phase D), depending on the level of maturity of the product. And processes 1–12 consist of cross-functional teams producing deliverables that are aligned by phase.

Figure 2-3 illustrates a tier 1 process that is integrated to external customer processes. In this framework, we see how internal milestones are aligned with the external customer milestones.

We treat managing engineering projects as a science. One of the most valuable principles of science is its structured approach to solving complex problems. Engineers are adept at breaking problems into logical components and then breaking each of these components into even smaller, more manageable pieces. This is called a work breakdown structure and is the key to breaking down a complete engineering project into tasks that can be understood, planned, estimated, scheduled,

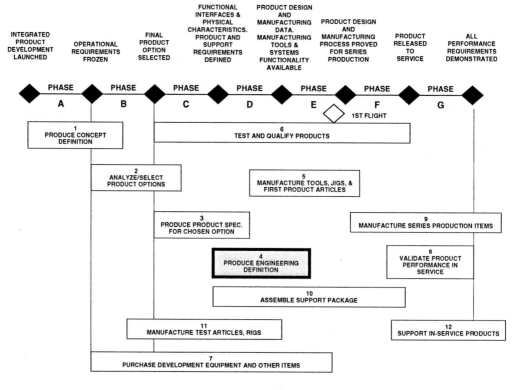

Figure 2-2: Process framework example 2 – Tier 1 processes (Courtesy British Aerospace Military Aircrafts and Aero Structures)

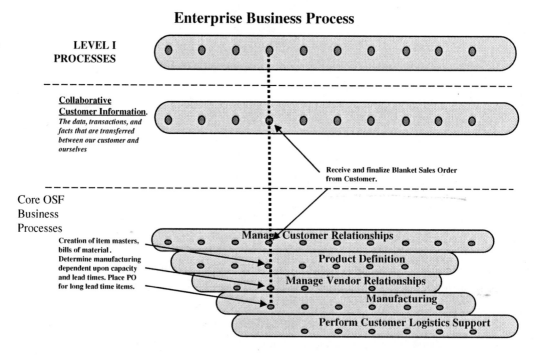

Figure 2-3: Process framework example 3 – Tier 1 (Courtesy Ontario Store Fixtures, Inc.)

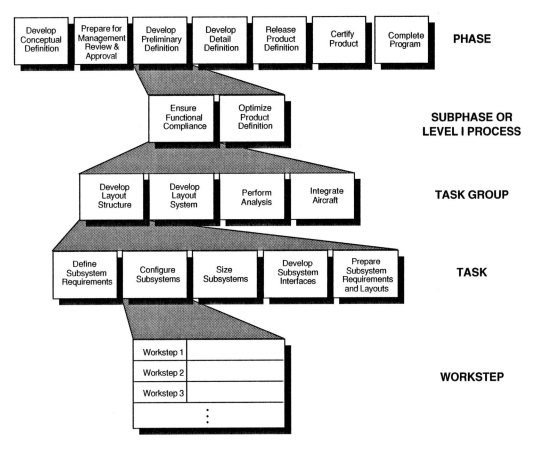

Figure 2-4: Work breakdown structure example

assigned, managed, tracked, controlled, and completed. We will discuss the process framework and its relationship to design quality project management and continuous improvement. As we will see later, the beauty of this framework is that good process management will deliver good project management on a silver platter.

2.2 WORK BREAKDOWN STRUCTURE

Phases

The first level of the work breakdown structure is the phase (see Figure 2-4). A phase can be decomposed into subphases, task groups, tasks, and finally worksteps. There is a logical relationship between each layer of the decomposition. A phase is a significant piece of work that can be planned, estimated, and managed as a complete project. Each phase has a budget, deliverables, and quality reviews. Phase boundaries provide limits for the work that is typically done between major management reviews and approvals.

Subphases and level I process

Each phase broadly categorizes related work, and the next level of detail, the subphase, defines a logical block of work that is smaller and more manageable. This block of work is documented in a workplan template. Subphases are often the starting point from which an engineering director tailors the templates to create a project workplan. Each subphase typically culminates in a formal review. Most subphases result in one or more customer deliverables. Many companies define these subphases as level I cross-functional processes that decompose into task groups and tasks. This is illustrated in Figure 2-2.

Task groups and tasks

Subphases are further divided into task groups and tasks. A task group is simply a logical grouping of tasks that facilitates organization of the planned work. An example of a task group in the manufacturing engineering area might be "develop assembly fixture." A task is a precise piece of work, with a detailed description, design tips, and examples.

Worksteps

Finally, each task is broken down into worksteps. A workstep is the most specific unit of work. An example of a workstep might be "calculate moment of inertia" or "design drill jig." It produces a single work product and can typically be assigned to a single individual. Each workstep should be documented with a concise description, estimating guidelines, the name of the resulting deliverable, and the recommended skill type for the person performing the work.

One company we studied has 7 phases, 26 subphases, 104 task groups, 520 tasks, and more than 5,000 worksteps in the process framework. The project management activities are similarly divided into subphases, task groups, tasks, and worksteps. This low level of detail, combined with the high-level structure, enables more effective project management and greater flexibility in the planning and execution of a project. The planning of the work effort can take place at any level, from phase to workstep, and the resultant workplan can be tailored to the specific needs of the project by modifying the templates.

Using a structured process framework to manage the business allows management to decompose strategic corporate goals to divisional, departmental, and individual goals. This is the foundation of a performance management system built around processes. A common thread links the top of the enterprise to the bottom and is illustrated in Figure 2-5.

Top management can use a process framework similar to that illustrated in Figure 2-6 to identify what their roles and responsibilities are to achieve the corporate strategy. Each element of cost, quality, schedule, and the like is addressed. For example, if we consider the define process, we see that product engineering is the key owner of this process. And the goals and objectives are to set cost and schedule objectives for engineering projects. This contrasts with division management,

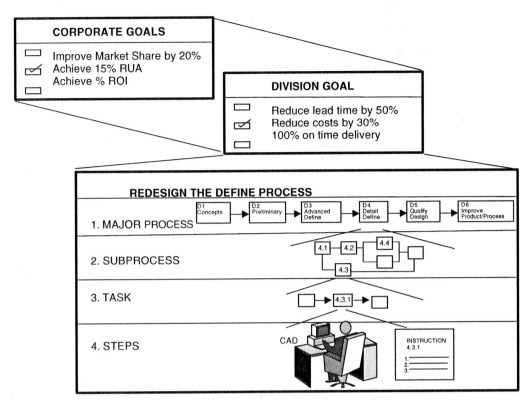

Figure 2-5: Management uses a process framework to link corporate goals to individual objectives

BUSINESS PROCESS	Key Owner Viewpoint	Goals and Objectives	Continued Operations	Profits	Quality	Schedule	Cost	Human Resources
STRATEGIC MANAGEMENT	Executive Corporate Management	Set these for the organization	Maintain markets	Be No.1 or 2 in each market segment	Customer satisfaction	Consistent with promises to customers	Lowest cost consistent with quality	Trained and capable workforce
OPERATIONAL & PROGRAM MANAGEMENT	Division Management	Meet and set goals while harmonizing with corporation goals	Maintain present level of operation; cont. quality improvement	Meet or exceed corporate goals	Meet or exceed own goals	Reduce cycle time	Cost reduction of at least 5% per year	Trained and capable workforce
ACQUIRE	Sales and Marketing	Meet sales plans and quotas	Maintain sales levels and revenues	Enough to meet sales incentive plans	No customer complaint; customer praise	No customer complaint; no problems with manufacturing	As low as possible to increase margins	Trained and capable workforce
DEFINE	Product Engineering	Meet, set goals, set project cost, and schedule objectives	Maintain existing product support and change management	Meet or exceed project objectives; set as design criteria	Integrate into design methods; meet project objectives	Employ cost-effective means to maintain schedules	Employ design methods to reduce product cost; use good management to reduce eng.cost	Empowerment; use human resource concepts during projects to max. extent
BUILD	Production Operations	Meet, set goals, costs,schedules and quality requirements	Using tracking systems, stay at least even, improve by 5%, more or less	Drive down costs, inventory work-in-process predistribution	Utilize advanced quality management processes to deliver quality	Consider whatever scheduling approach that works	Continual quality improvement (CQI) to improve process and product cost	Trained and capable workforce; improve labor relations
SUPPORT	Customer Service Management	Meet and set goals while harmonizing with corporate goals	Meet sales plans and quotas	Meet own goals while supporting other organizational goals	Push quality into organization	Meet customer promises	Cost reductions when consistent with promises	Support training

Figure 2-6: Business process vs. company business objectives matrix

Type	Characteristic	User	Relationship to the Business	Comments
Milestone	Major events	Executive	Goal achievement and revenue/cost/ profitability	Visibility provides opportunity to manage by indicators; helps to understand variation
Schedule	Task status and completion monitoring (when)	Director	Event monitoring; early/late - budget variance	Provides opportunity to intervene to provide leadership, by anticipating variation
Business Process	Process descriptions (how, why)	Manager	Process monitoring; priorities; resources; progress; deliverables completion; next steps	Provides details on how and why something should be done; uses schedule to monitor process activities via indicators
Routing	People std time, tools, physical things	Manager / Supervisor	Physical version of manufacturing process	Applies process description to actual use
Activity	Task descriptions (what, who)	Supervisor	Task completion focus; priorities; steps or tasks to complete	Provides details on what activity is to be accomplished
Work Center	Where work is performed	Supervisor / individual	Physical version of activity	Applies activity detail to a specific work center or other physical location

Figure 2-7: Process information map to user

which owns the operational and program management process. Their goals and objectives cross the whole organization and must be harmonized with corporate goals. We can see the viewpoint of each process owner.

In a process-oriented environment, the types of information and the individual interests vary by organizational responsibility. This is illustrated in Figure 2-7. For example, we see that milestones are used by executive management to track major events. The relationship to the business is to achieve revenue goals, costs, and profitability. This contrasts with routing information that is used by first-line management and supervisors to manage the manufacturing process.

When you use process management techniques, business processes are conceived, designed, and executed at the appropriate levels. The concept of design and manufacturing engineering in a concurrent environment showing the deliverables is outlined in Figure 2-8. The first step is to gain agreement on these workflows and then to develop the relationship between deliverables.

In Figure 2-8, we can see that the Acquire process produces the customer requirements which are used by the Define process to create work statements, bills of material, and the like. And these deliverables are used by the Build process to create material requirements. The logical relationship of deliverables create workflow.

2.3   CUSTOMER DELIVERABLES

Just as the process framework defines a work breakdown structure for program activity, the resulting deliverables are structured to organize how we create value. Customer deliverables are the main focus of the product development effort, but they are actually created as a natural consequence of the work performed during a

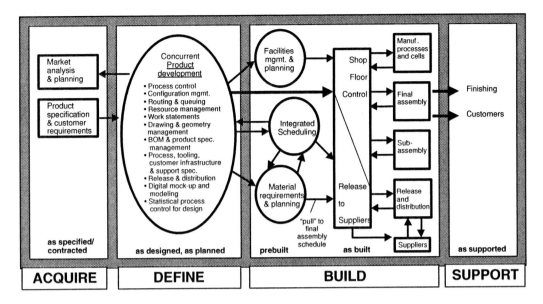

Figure 2-8: Define/Build interface

program. It is through these deliverables that we create value and therefore profit. Understanding the concept of the "value of knowledge" is simpler in today's connected world, where companies collaborate to complete designs. However, as recently as the early 1990s, it was often difficult for engineers to accept the concept, for example, that a completed stress analysis had economic value.

Just as the process architecture structures the activities of the enterprise, the deliverables architecture structures the outputs of those processes. Figure 2-9 illustrates the concept of a deliverables architecture. Figure 2-10 provides an example.

The lowest level of deliverable is the work product, which results from the successful completion of a workstep. An example of a work product might be an engineering drawing or a manufacturing method sheet for a component. Some work products are merely interim deliverables, created for use in other worksteps, and are retained as working papers in the project repository. If the company operates in a virtual environment and does not produce paperwork, solid models and engineering data are stored in the CAD system or PDM system (see Chapter 11). Other, more significant work products are combined to form key deliverables. A *key deliverable* is a logical grouping of related work products, which is reviewed by the IPT for completeness and accuracy. Related key deliverables are then assembled into a customer deliverable, which is presented to the customer for approval. All work products, key deliverables, and customer deliverables are also retained as working papers for the project.

The structure of these deliverables, shown in Figure 2-9, is one of many but serves to illustrate the hierarchical relationship between work products. Once the

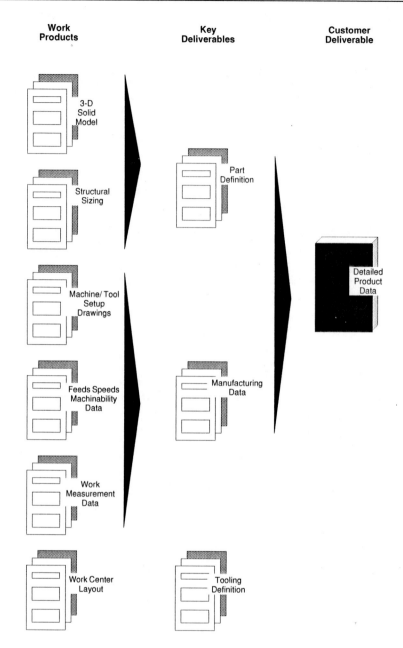

Figure 2-9: Deliverables architecture

logical relationship of the deliverables has been defined, the business processes can then be mapped. In Figure 2-9, we can see that detailed product data, the high-level customer deliverable, consists of three key deliverables – tooling definition, manufacturing data, and part definition.

Figure 2-11 identifies a set of standard, formal outputs, or customer deliverables for each of the core business processes. These deliverables are presented at

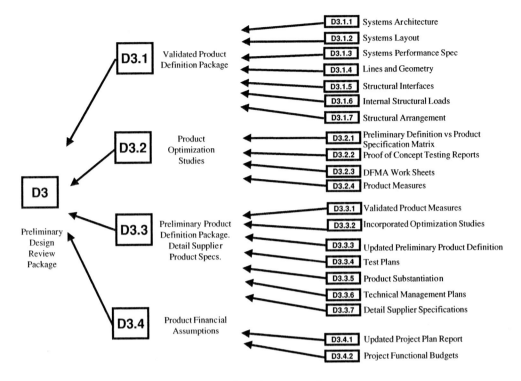

Figure 2-10: Deliverables architecture flow chart example

specific points in the project, after major pieces of work have been completed. Deliverables indicate that a task has been completed. They also provide necessary input that allows the next task to be completed. Figure 2-11 shows a high-level example of customer deliverables in the process framework.

Table 2-3 (see Appendix B) provides a brief description of each of the customer deliverables defined by the process framework at the phase level, and Table 2-4 (see Appendix B) provides a brief description of each of the customer deliverables at the subphase level.

Within the structure of the process framework, the data defining the product are complex and interrelated. To make sense of the data and make it accessible to everyone, a simple structure must be developed. That structure must relate the part of the physical product and the type of requirement placed upon it (see Section 3.1, "Single Number Tracking System").

2.4   MILESTONES AND MATURITY GATES

The concept of product maturity gates is vital to managing the Define process. A maturity gate is often referred to as a milestone, but the idea of a product maturing and gaining certain rights of passage is an important one. The program director controls passage through the maturity gates. Their experience from many previous

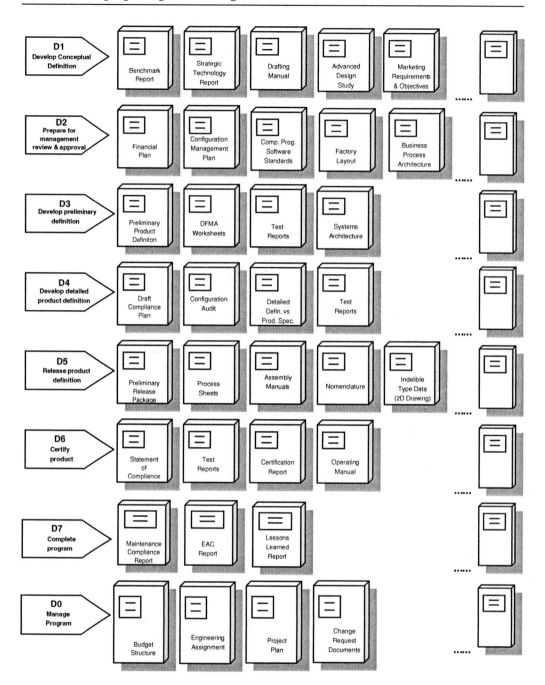

Figure 2-11: Customer deliverables mapped by process framework example

programs at similar review points provides a powerful basis of comparison and quality control. We lay out clear expectations and, therefore, provide a simple basis of comparison. Design alternatives occur at each milestone. Decisions as to which alternative to choose are made. This is the process of convergence illustrated in Figure 2-12.

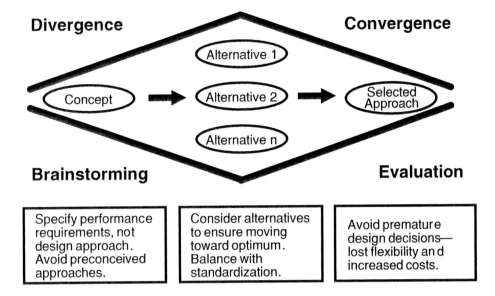

Figure 2-12: Design alternatives

Figure 2-13 illustrates major milestones within and between phases in the product development life cycle.

Before the product can pass through the maturity gate, the IPT prepares a predefined set of deliverables. Table 2-5 (see Appendix B) illustrates some of these examples.

The phase/maturity alignment matrix enables the relevant disciplines and functional area to align and integrate their processes with the deliverables. This is illustrated in Figure 2-14. We can see that vehicle systems must produce the system location and routing deliverable to move through maturity gate D2. In this matrix, product maturity is linked to subphase gates, and each discipline has deliverables defined relative to each product maturity subphase. These deliverables are the enablers in successfully reaching a consistent level of product maturity.

The subphases identified are a generic set, identified to control product maturity. The subphases are not mandatory so that where the program requires more or fewer phases, the change in their quality/program plan must be documented. However, the concept of maturity gates is central to managing the engineering process, and exit criteria from each gate should be documented and strictly enforced.

The deliverables constitute the exit criteria from the current subphase, and the input criteria to the following subphase. Figure 2-15 illustrates an example of detailed deliverables by function and maturity gate. We can see that the vehicle systems function must produce eight deliverables to move through maturity gate M1. And these must be aligned with the deliverables created by the other functional disciplines. Allowing the project to pass through a gate with some deliverables still incomplete may serve to move the project ahead, but it also introduces an element of risk.

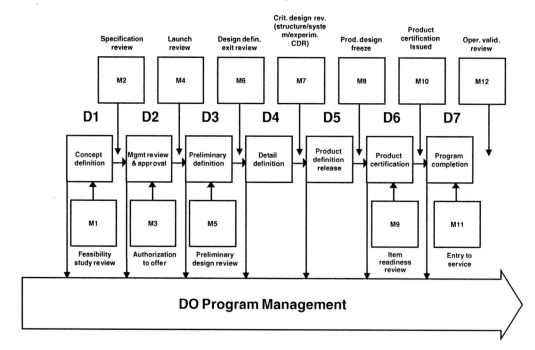

Figure 2-13: Maturity gates and the process framework

Functional Interfaces & Physical Charactetistics,
Product & Support Requirements Defined

Product Design & Manufacturing Data,
Manufacturing Tools & Systems Functionally Available

	Sub Phase D1.1	Maturity Gate D1	Sub Phase D1.2	Maturity Gate D2	Phase D1.3	Maturity Gate D3	Phase D1.4	Maturity Gate D4	Phase D1.5	Maturity Gate D5	Phase D1.6
Design Structures & Manufactur'g	Concept Options Phase	Concept Options Frozen	Concept Definition Phase	Concept Freeze	Detail Options Phase	Detail Options Frozen	Detail Design Definition Phase	3D Design Freeze	Full Design Definition Phase	Product Design Freeze	Manufg Definition Phase
Vehicle Systems	System Requirem'ts Verification Phase	System Spec's Issued	System Installation Options Phase	System Location & Routing Defined	Detail Physical Design Definition Phase	Detail Physical Design Definition Phase	Physical Design Verification Phase	Installed System Performance Verified (CDR)	Funct'nal Design Validation Phase	Systems Functions Validated	Testing Definition Phase
Software	Software Requirem'ts Analysis Phase	Allocated Baseline	Preliminary Design Phase			Preliminary Design Review	Detailed Design Phase	Critical Design Review	Coding & Unit Testing Phase		
Equipment Engineering	Eqpmt Contract Specificat'n Phase-high Risk Eqpmt	Eqpmt Contract Awarded High Risk Eqpmt	Eqpmt Contract Specificat'n Phase-low Risk eqpmt	Eqpmt Contract Awarded Low Risk Eqpmt	Preliminary Design Phase	Preliminary Design Review	Detailed Design Phase	Critical Design Review	'A' Model develop'nt Phase	'A' Model Delivered	'B' Model develop'nt Phase
Electronic Engineering	Specificat'n Analysis Phase	Specificat'n Agreed	System Design Phase			System Design Frozen	Detailed Design Phase	Detailed Design Frozen	Manufacture And Test Phase	Validation	Environm't & EMC Testing Phase
Aerospace Ground Eqpmt Engineering	Wstge Definition Phase	Wstge Requirem'ts Defined	Wstge Planning Phase	Wstge Support Planning Complete	Wstge Requirem'ts Definition phase	Age Requirem'ts Defined	Age Tender Phase	Age Tender Activities Complete	Age PDR/CDR Phase	PDR/CDR Activities Complete	Age Full Defn & final Maintenance Planning Phase
Reliability Attainability Testability	Detailed RM&T Requirem'ts Definition Phase	Detailed Requirem'ts Defined	Preliminary RM&T Analysis Phase			Compliancy Illustrated at PDR	Final RM&T Analysis Phase	Compliancy Assured At PDR	Detailed Logistics Requirem'ts Definition Phase	Data Analysis Complete	

Figure 2-14: Gate alignment matrix

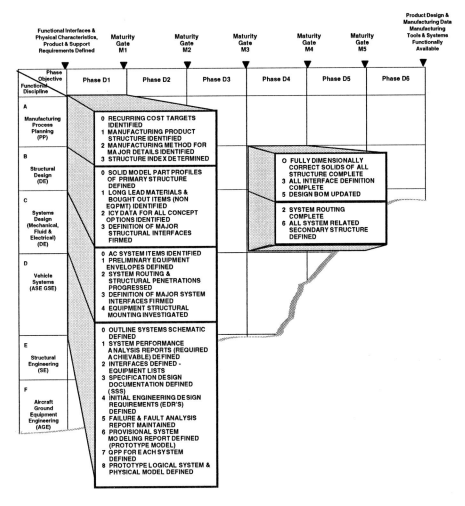

Figure 2-15: Maturity matrix example (Courtesy British Aerospace Military Aircraft)

In applying the engineering process generically, individuals need to know what is expected of them within the new environment. The matrix identifies the deliverables required from each discipline relative to product maturity subphases. The matrix clearly identifies an integrated product development approach by looking down the vertical columns.

2.5 PROCESS MATURITY

Process management can simply be implemented as a methodology in and of itself. However, acting holistically, the idea is to integrate previously distinct bodies of knowledge. Making this change is itself a process. Good process management requires that we measure the process performance. In addition to measuring design quality, cost, and schedule, measuring process maturity will show how

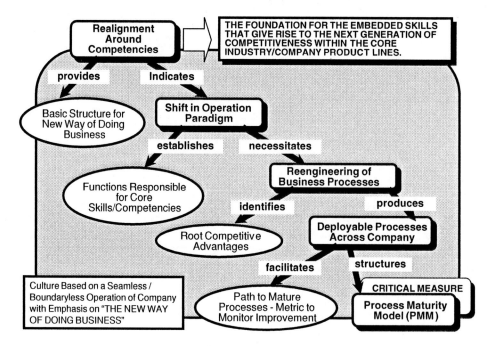

Figure 2-16: Process maturity route map

the define process compares with world-class benchmarks. In a collaborative-commerce project, discussing with partners the progress of the Define process implementation will provide valuable insight. Processes must be mature to ensure that deliverables are mature. Figure 2-16 illustrates the major tasks required to implement a formal process maturity mode. Unfortunately most companies give up at reengineering business processes in isolated functional areas of the company. Please refer to the process maturity self evaluation tools in Appendix A.

The five-level maturation concept shown in Figure 2-17 is adaptable to all processes in the enterprise and in particular to product development. It illustrates the process performance levels that a company must achieve to gain world-class maturity.

Figure 2-18 illustrates the detailed criteria that must be met to move from one level of maturity to a higher level. A description of the requirements at each level follows.

Level 1: Baseline process established (reengineer and document)

- ○ Process is identified and defined. Mapping is available.
- ○ Process owner is established.
- ○ Both inter- and intrafaces have been established and documented through coordination meetings. Interfaces are under control and managed.
- ○ Activities for processes and subprocesses are documented and have entry and exit criteria/points.

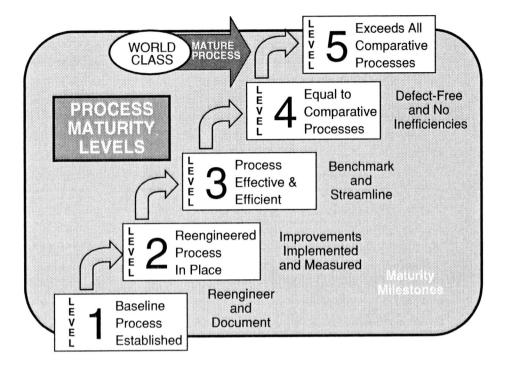

Figure 2-17: Five-level process maturity model

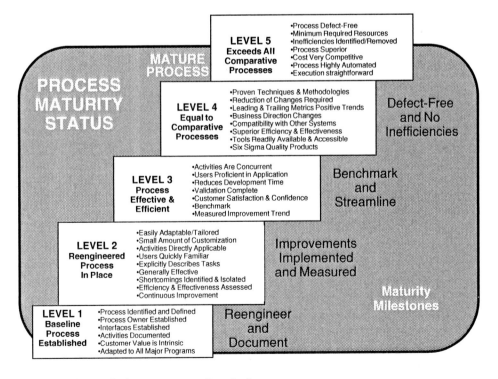

Figure 2-18: Detailed process maturity criteria

○ Customer value is intrinsic in the process, and the customer agrees that the process applies to his/her program.

○ Process is adapted to all major programs, and adaptation incorporates resolution of deficiencies and redundancies.

Level 2: Reengineered process in place (improvements implemented and measured)

○ Customer agrees the process is easily adaptable/tailorable to meet the needs of the program.

○ Process requires only a small amount of customization to meet project/program (application) needs/requirements.

○ Process is centered on activities directly applicable to the program and provides information necessary to complete work packages.

○ Process is engineered to allow users to become familiar quickly with its application, progressing along a very steep learning curve.

○ From a customer's viewpoint, the process explicitly describes the activities, the prerequisites, expected result, competencies (types of people), techniques/methodologies, and mechanisms involved with the process.

○ Deployed process is generally effective as viewed by the customer.

○ Shortcomings are identified, and the root cause is isolated (deficiencies and redundancies).

○ Efficiency and effectiveness assessments are established in relation to the customer's expectations/needs/requirements.

○ Metrics are established for feedback mechanisms for continuous improvement.

Level 3: Process effective and efficient (benchmark and streamline)

○ Activities are concurrent where practical to allow parallel accomplishment and initiation at the earliest possible time.

○ Process users are proficient in the application of the process to develop products for the programs, from the customer's viewpoint.

○ Process significantly reduces product development time, cycle time, and iterations to produce the required result/product (measurable).

○ Process dramatically reduces the cost for development and producibility.

○ Validation of the process is complete, and the risks associated with the process are defined and understood with workable abatements in place.

○ Customer's satisfaction and confidence in the process and its application are very high.

○ Benchmark testing is conducted, and the process is rated well within the category of the best 10 percent (effect as measured by the benchmark results).

○ Measures (metrics) show an improvement trend as a result of the process improvement actions in work.

Level 4: Generally exceeds comparative processes (defect-free and no inefficiencies)

- ○ The latest proven techniques and methodologies are incorporated into the process to enhance efficiency and effectiveness.
- ○ Objective evidence is available showing the effect the process has had on the reduction of changes required during development and production (i.e., engineering change notices – ECNs are significantly reduced).
- ○ Both leading and trailing metrics are continuing to show positive trends to ensure institutionalization of the process.
- ○ Process provides information contributing to predictable business direction changes and meets the customer's requirements for future competitive proposals.
- ○ Compatibility of process with other systems in the company is established and contributes significantly to the business posture of the company.
- ○ Few processes involved in the benchmark activities (similar in scope, etc.) are superior to the process in efficiency and effectiveness.
- ○ All tools used by the process are readily available and accessible to all users with detailed instruction on application of tools and point in the process where tools are used.
- ○ Process produces Six Sigma Quality products as viewed by the customer. Six Sigma Quality is a standard whereby the standard deviation of actual outputs of a process (Sigma) is so small that even a range of plus/minus three Sigma is within acceptable tolerances.

Level 5: Exceeds all comparative processes (mature process)

- ○ Effectiveness measures indicate that the process is defect-free from the control viewpoint, and the customer agrees with these data.
- ○ Process is executed with a minimum of required resources, and the cycle time is equal to or less than the best process considered in the benchmarking conducted.
- ○ Inefficiencies are identified and removed from the process, and the customer agrees with this result.
- ○ The cost of using the process is very competitive, and the process produces a quality product.
- ○ Execution of the process is straightforward and can be learned very quickly.
- ○ The process is highly automated, and the tools employed have user-friendly interfaces.

It is absolutely necessary for your company to implement the process maturity model to maintain world-class performance. Many companies spend millions of dollars to design and create a process integration framework, but they stop after the process framework is documented. Process must be measured and continuously improved.

3 Organization of Engineering Tasks

This chapter explains the organization of tasks using workplan templates within the task guidelines (see Section 6.2). Within this structure, you will be able to see how a task fits with your program plan and to know how to find detailed information about it.

Tasks depend upon two other elements – resources and schedule. Understanding the resources required to perform a task and the schedule within which the task must fit allows us to manage costs effectively. All three elements must be managed in concert with one another. A change in one element has an impact on the others. Prior to a formal process management, these elements, in many organizations, were managed independently. This imbalance would lead frequently to internal power struggles. Functional directors would assign resources as they saw fit without a view of the impact on other tasks within the integrated master schedule.

Tasks are always organized into the processes that define the business process framework, and processes overlap phases. In addition, developing reference manuals that capture the knowledge you gained on previous programs and that document new or established approaches offers tremendous benefits. They describe the worksteps of how to carry out a task, including technical rules. The documentation of these rules allows the organization to "learn." The process of maintaining design rules lends itself well to intranet application because it allows distributed access to all members of the collaborative project. Further, an intranet permits ease of update and version control. (See Chapter 11 on utilizing product data management.)

An engineering task describes a discrete piece of work to be done on an engineering design project and is organized like a database. We can access data about a particular task by using a tracking system.

3.1 SINGLE-NUMBER TRACKING SYSTEM

Single-number tracking in a structured system organizes the planning documents of a program. In later chapters, we will discuss the specification tree, work breakdown structure, statement of work, integrated master plan, and integrated master schedule for the program. They will all be defined by a common numbering

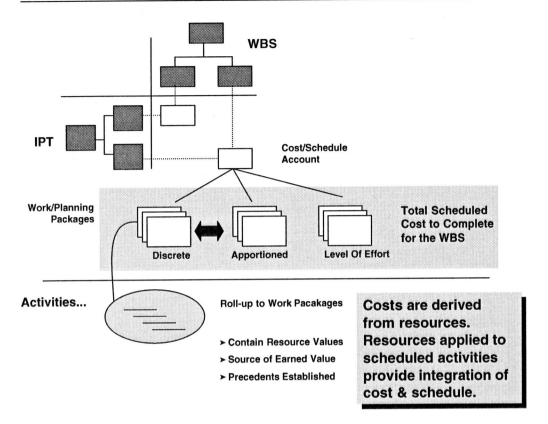

Figure 3-1: Specification tree

system. This numbering system allows the integrated product teams to trace document links forward or backward to retrieve the information from these planning documents quickly. This numbering scheme is subsequently used on all future planning documents to create a trail that provides accountability, traceability, and reference points for each task. This single-numbering system provides an excellent hook to the cost accounting/earned value systems. Therefore, the actual numbering convention used throughout the documentation should be compatible with your specific system. The process begins by developing the requirements specification tree, which is typically broken down by product, subsystem, or service. For each element on the specification tree, a WBS element is defined (see Figure 3-1).

The specification tree and WBS then share a common numbering system and are both product focused. Tasks that go across the entire program are rolled up under their own WBS element. Next, statements of work are typically written for each WBS element, and they too employ the common numbering system used on the specification tree and WBS. At this point, an integrated master plan is constructed as an extension of the SOW. It contains events, significant accomplishments, and associated accomplishment criteria as outlined previously. Once again, the IMP shares the common numbering system with the specification tree,

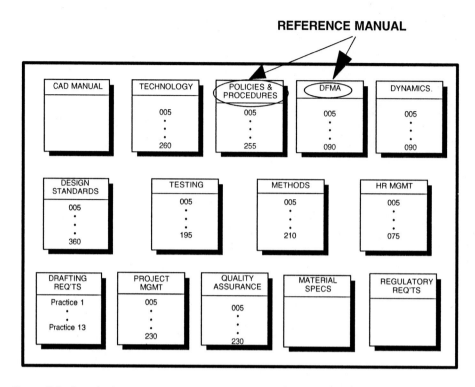

Figure 3-2: Standards, specs, and references organized by technical discipline

WBS, and SOWs. Finally, the detailed tasks necessary to accomplish the work defined in the IMP are put on a time line in the form of an integrated master schedule. For a detailed description of how to create an IMS, see Chapter 7. These tasks are also linked to the numbering scheme originated in the specification tree development. The purpose of this numbering exercise is to structure a process involving all program team members and support functions for putting together an integrated plan and schedule for the program. The process helps the team embrace redefined processes at the earliest and most meaningful time in the program. By using a common numbering system and imparting a product focus into the specification tree, WBS, SOW, IMP, and IMS, the customer and contractor gain increased visibility and accountability to the product cost and program performance.

Just as product-related data have a single-number tracking system, the define process documentation is organized by the technical discipline. Analysis, design, quality assurance, and testing documentation is organized using a structured numbering system. In Figure 3-2, we see that each task is linked to a reference manual, which contains information to help team members complete a task. In Figure 3-2, we highlight the policies and procedures manual and the design for manufacturing and assembly (DFMA) manual.

Within each workplan template, a task is given a process framework number to indicate its relative position in the process framework. Then we add a reference to the appropriate manual. Figure 3-3 illustrates task D4.1, which is linked to

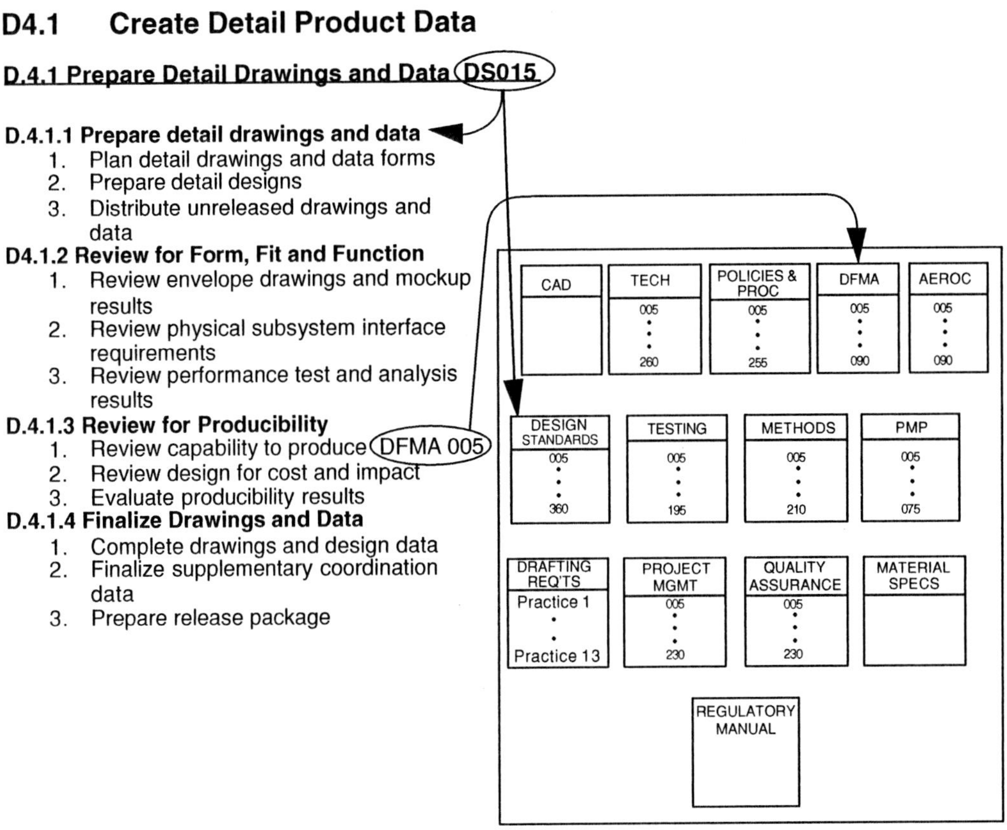
D4.1 Create Detail Product Data
D.4.1 Prepare Detail Drawings and Data DS015
D.4.1.1 Prepare detail drawings and data
1. Plan detail drawings and data forms
2. Prepare detail designs
3. Distribute unreleased drawings and data
D4.1.2 Review for Form, Fit and Function
1. Review envelope drawings and mockup results
2. Review physical subsystem interface requirements
3. Review performance test and analysis results
D.4.1.3 Review for Producibility
1. Review capability to produce DFMA 005
2. Review design for cost and impact
3. Evaluate producibility results
D.4.1.4 Finalize Drawings and Data
1. Complete drawings and design data
2. Finalize supplementary coordination data
3. Prepare release package
CAD
TECH 005 260
POLICIES & PROC 005 255
DFMA 005 090
AEROC 005 090
DESIGN STANDARDS 005 360
TESTING 005 195
METHODS 005 210
PMP 005 075
DRAFTING REQ'TS Practice 1 Practice 13
PROJECT MGMT 005 230
QUALITY ASSURANCE 005 230
MATERIAL SPECS
REGULATORY MANUAL

Figure 3-3: Engineering tasks organized by business process number

design standard DS015. DS015 might relate to the application of tolerances to drawings. Similarly task D4.1.3.1 relates to DFMA 005. DFMA 005 might relate to producing sheet metal components on a break press.

3.2 INTEGRATED MASTER PLAN

The program's integration team constructs an integrated master plan after the WBS is established. It involves building an event-driven plan that describes the activities necessary to produce and deliver a product that meets contract requirements (see Table 3-1, Appendix C). The IMP defines the total program by identifying the maturity gates (see Section 2.4) and then defining criteria that permit passage. The IMP milestones are sequenced into an integrated process hierarchy (Figure 3-4) and then supplemented with narratives for detailing the key processes and tasks necessary to meet the contractual requirements. The IMP also shares the common numbering system with the specification tree and WBS.

Customer tailoring or contract-unique requirements must be incorporated into the IMP. These may take the form of narratives. In addition, depending on the

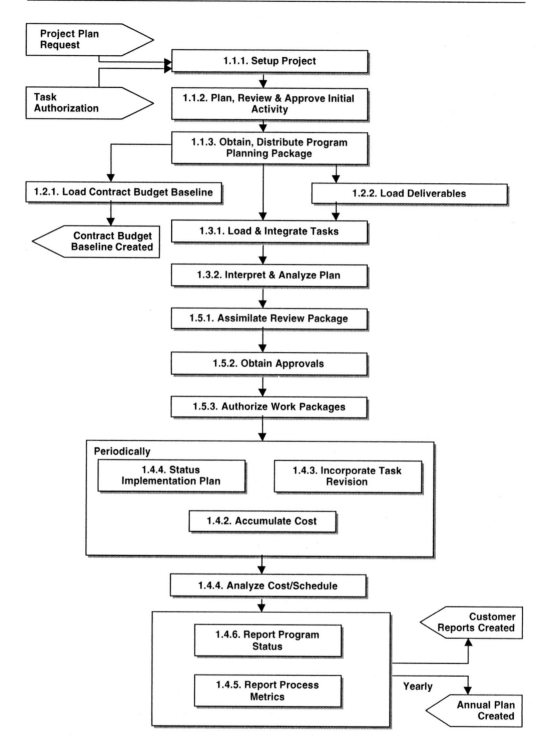

Figure 3-4: Task planning process data flow diagram – Tier 3 of process hierarchy

Workplan Template

Unit Description	No. of Units	Estimating Guidelines	Hrs./ Unit	Per Hours	Assigned To	Skill Type

D4.2 Confirm Detail Configuration by Analysis

Process #

D.4.2.1 Develop Design Layout

D.4.2.1 Review Preliminary Design (DS005)

Task	No. of Units	Estimating Guidelines	Hrs./ Unit	Per Hours	Assigned To	Skill Type
1. Identify preliminary design problem areas	—	1-3-5	—		—	DES
2. Identify information required	—	0.5-1-2	—		—	DES
3. Gather/create information	—	1-3-5	—		—	DES
4. Evaluate design alternatives EPP Ref. #	—	0.5-1-4	—		—	DES

D.4.2.2 Analyze and Refine Preliminary Design Layouts (DS025)

1. Coordinate and analyze data	—	2-8-16	—		—	DES
2. Design layouts and alternates	—	0.5-1-2	—		—	DES
3. Evaluate improved layouts and alternates	—	1-2-4	—		—	DES

D.4.2.3 Layout Interfaces (IN060)

1. Collect and analyze data	—	1-2-4	—			
2. Determine interface types and locations	—	1-3-5	—		—	DES
3. Develop interface layouts	—	2-8-40	—		—	DES
4. Evaluate and refine interface layouts	—	1-5-8	—		—	DES

D.4.2.4 Analyze and Refine for Technological Impact (AN005)

					—	DES
					—	DES
1. Collect and analyze data	—	3-5-8	—			
2. Perform external systems analyses	—	1-3-5	—		—	DES
3. Perform internal systems analyses	—	1-3-5-	—		—	DES
4. Perform interface analysis	—	1-3-5	—		—	DES
5. Evaluate analyses	—	0.5-1-2	—		—	DES

D.4.2.5 Prepare Design Layouts (DS010)

1. Prepare basic layouts	—	1-2-4	—		—	DES
2. Prepare interface layouts	—	1-3-5	—		—	DES
3. Evaluate layouts	—	2-8-40	—		—	DES

Figure 3-5: A workplan template

phase of the program (i.e., whether proposal or detailed definition), it may contain unique short-term events or tasks defined by the integrated product teams. The integration product team or program manager will monitor these tasks to assess program status. In a product development program, the IMP serves as the top-level plan, incorporating all major customer and program milestones. As such, subsequent detailed plans assembled by each IPT become the plans and schedules to which these teams are measured during the program.

3.3   INTEGRATED MASTER SCHEDULE

Next in this process of integrating resources, costs, and schedules, the detailed tasks in the integrated master plan (IMP) are put on a time line in the form of an integrated master schedule. The IMSserves as an interteam schedule. The IMS should be used as the primary tool to assess program schedule performance, evaluate critical path, schedule margin and risk, and perform what–if analyses. At a minimum, the IMS should include traceability of key tasks to all accomplishments submitted in the IMP. Lastly, it is a recognized best practice to establish an acceptable maximum percentage of program schedule deviation to be managed by the program integration team.

3.4 DEVELOPING A WORKPLAN TEMPLATE

Rather than linking the reference materials for tasks to the number in the process framework, which could change over time, we suggest linking them to "absolute" engineering policy and procedure (EPP) numbers that will never change. So that you can access task guidelines, the project plan lists both the process framework number and the EPP reference number. The same EPP reference number will always remain attached to the same task name, even if the project plan changes. This is illustrated in Figure 3-5.

Figure 3-5 also shows other information that can be included in a workplan template. This information includes the time estimate to perform a task, the skill level, and the name of the assigned person.

The EPP reference number is an important key to finding information about engineering tasks. Not only will you use it to access task guidelines, but you will also use it to identify sections of deliverables related to your task.

Chapters 6 and 7 explain in detail how to use these numbers, for the day-to-day work of a project.

Part 2

Applying Engineering Processes to Program Management

Roles and Responsibilities

The successful implementation of an advanced engineering program is a team effort. To be effective, the program team must be more than just a collection of individuals. The team must have the management, technical, functional, and product skills necessary to address the variety of challenges typically encountered on an advanced engineering program. Moreover, your company's partners and subcontractors must work well together to provide an effective solution for the customer.

As organizations grow larger, hierarchical structures, specialization, and facility dispersion make it more difficult to communicate across functional boundaries and to coordinate activities. Integrated product teams are a way to organize into smaller units to facilitate communication among customer, suppliers, partners, and your company; to develop a multifunctional perspective; and to develop well-balanced, consensus decision making.

The organization of the team members on the program is a critical aspect of the proposal and program initiation processes. Before the program can get under way, it is necessary to identify the major program roles and responsibilities. Individuals should then be selected to fill those roles based on their skills and experience in solving similar problems.

Customer personnel, operational/program management, and partners play specific roles in the management and execution of an engineering program. The organization for a specific program depends on the product design, the particular program responsibilities of your company, customer and partners, and other program characteristics. However, certain roles are common to every program. The following discussion considers these roles, and Figure 4-8, at the end of this chapter, shows how they fit into a sample program organization.

4.1 THE CUSTOMER

The following paragraphs explain the roles and responsibilities of a customer's involvement in a program, also included is a paragraph regarding the critical success factors of customer involvement. For a quick reference, Tables 4-1 through 4-3 show the responsibilities in relation to the subphases within program management

Table 4-1: Proposal phase roles and responsibilities

Task	Program management	Integration team	Integrated product team	Functional management
Proposal responsibility	Ensures IPD methodology on proposal through IPD methodology book and integration team involvement.			Ensures that functional policies and procedures are followed through process auditing.
Organization		Identifies requirements for multifunctional proposal teams in conjunction with product-focused WBS.		Supplies resources based on program requirements to generate product-focused technical approach, cost, and integrated schedules. Functional process, metrics, and tools are preapproved by the functional manager.
Requirements	Identifies design-to-cost/life cycle cost targets.	Establishes WBS and systems interface plans in conjunction with IPD methodology guide. Identifies cost account levels and appropriate shop order levels. Identifies customer requirements and flowdown to IPTs.		
Estimating	Allocates design-to-cost targets Generates colocation plan	Upon approval, the proposal team integrates and may tailor functional inputs to establish multifunctional needs focusing on high-quality products to meet customer requirements. Any metrics/estimates that are tailored are forwarded to the respective function. The proposal team must develop mitigation plans associated with the functional risk. The proposal team upon completion of the integration process is responsible for the content of the proposal. Responsible for functional concurrence of risk mitigation plan or consensus at the next level of management.	The proposal team reviews and concurs with functional metrics/estimates and inputs on technical approach, DTC targets, and schedule. Inputs proposal assumptions and risks into database. Reviews database periodically for proposal and program impact, update as required.	Supplies approved functional metrics. Approves internal estimating tools. Reviews metrics/estimate tailoring and identifies associated risk. Concurs with risk mitigation, or provides program issues/reasons for nonconcurrence. Review database periodically for functional impact, update as required.
Review and approval	Reviews and approves technical approach prior to internal design reviews. Chairs multifunctional product-oriented proposal cost reviews. Approves proposal.			Supports review of technical approach. Participates in multifunctional product-oriented proposal cost reviews. Approves proposal.
Negotiations	Program director participates in and/or influences negotiations with customer (prior to contract award as well as during program life cycle). Proposal manager reviews and submits SOW changes from negotiations to proposal team. Proposal manager updates design-to-cost baseline from negotiations.			

Table 4-2: Program start-up roles and responsibilities

Task	Program management	Integration team	Integrated product team	Functional management
Proposal responsibility	Ensures holistic approach on proposal team involvement. Program director ensures program is planned in manner compliant with all policies and procedures. Program director establishes management reserve for program.	Program director and IT direct program and delegate management and budget authority to appropriate IPT levels in conjunction with DTC targets and risk. IT generates and is responsible for IMP and IMS. IT flows proposal plans, budgets, risks, and assumptions to IPTs.		Program director and company president consultant on technical program issues. Ensures that functional policies and procedures are followed through process auditing.
Program staffing		IT and IPT leaders are responsible for manpower phasing and concur with functional staffing plan. IT conducts multifunctional IPT plan reviews.	IT and IPT leaders are responsible for manpower phasing and concur with functional staffing plan. Establishes template for skill set of IT and IPTs required to support the program; updates as required. IPTs generate and are responsible for detailed product development plans identifying product requirements and objectives including performance measurements, integrated plans, and integrated schedules with milestones. Participate in multifunctional product-oriented IPT plan reviews. IPT's identify additional issues, risks and assumptions on negotiated scopes, budgets and schedules	Provides staffing and hiring per business needs. Assigns and reassigns team members as required in concert with program IPT leaders. Assigns IT members and IPT leaders in conjunction with defined skill sets and in accordance with program manager. Review requirements flow down as appropriate for detail plan approval.
Functional process tailoring		IT documents process tailoring and tools and technologies plans in program directive.	IPTs identify functional process tailoring and tools and technologies needs. Is responsible for risk mitigation plans. Generates program review plan.	Concurs with or identifies risk in functional process tailoring and tools and technologies needs. Approves of risk mitigation plan or provides program issues/reasons for nonconcurrence.
Colocation		Updates and implements colocation plan.		
Training	Identifies with IT and IPT's training needs.			Facilitates training.

Table 4-3: Program implementation phase roles and responsibilities

Task	Program management	Integration team	Integrated product team	Functional management
IPT implementation			All team members responsible for technical, budget, schedule and quality through program life cycle. Technical, cost, and schedule priorities are determined by IPTs and approved by IT and program director.	Provide functional resources, processes, and tools to program to meet program requirements. Functional process director responsible for process, training, and tool administration and continuous process improvement.
Program review reporting	Program director conducts program reviews focusing on cost, schedule, and technical performance. Progress of future tasks, events, and milestones will be reviewed.	IT approves all IPT work authorizations and changes in work packages. IT communicates new major program milestones or change in program requirements. IT acts as requirements change control board. IT ensures that best practices, processes and methods are applied to the program needs and in conjunction with the handbook.	Identifies performance against DTC targets. Identifies performance against functional metrics. Risk and mitigation plans. IPTs submit cost and schedule variances with corrective actions to IT for review and approval. IT and IPTs hold meetings with team member. IPTs monitor DTC targets from proposal.	Participates in multifunctional reviews and provides feedback to program director. Provides functional best practices, processes, tools, analysis techniques, etc. Provides support to programs in resolution of technical issues as requested by IT. Reviews variances at request of program director. First-level director provides feedback to program through consultation of assigned functional members. Develops and documents best practices. Assists IT and IPTs in technology selection consistent with DTC goals and functional capabilities.
Customer interface	Program director provides interface to customer and establishes lines of communication. Program director chairs program reviews with customer.		Supplies variance reports to contracts.	Contracts supplies variances reports to customer.
Metrics			Develops programmatic metrics implementation team. Provides input on functional metrics realized on program per standard functional format.	Develops functional metrics and formats and continues process improvements.

Design reviews	Coordinates trade study process among IPTs and approves trade study plans and budgets.	Presents technical approach at customer design reviews. Performs supplier design reviews. Identifies action items, risk, and mitigation plans. IPT leaders coordinate reviews per IMP and IMS. Schedules and performs internal design reviews. IPTs conduct trade-off analysis. Presents technical performance. Responds to action items. Identifies performance against DTC targets. Identifies performance against functional metrics.	Reviews program design review package prior to customer review. Supplies technical expert outside of program to participate in supplier design reviews at request of program manager. Selects personnel to support internal design reviews, and concurs with action item responses. Provides appropriate design review process and checklists of items to be reviewed at design reviews.
Lessons learned		Provides lessons learned on process improvements to function. Provides feedback on process improvements operations lessons learned database	Provides appropriate functional lessons learned to program. Institutionalizes lessons learned in functional process. Participates in review of lessons learned.
Plan and equipment		IPTs define P&E requirements to support program needs. IPTs provide input to P&E for future business needs.	Implements P&E requirements to meet productivity goals and supports functional and program needs.
Staffing	Program director and functional management identify appropriate IT members for program.	Together with functional management ensures program IPTs have the appropriate/adequate personnel and resources to perform program goals. Provides functional organizations with program manpower phasing updates throughout program.	Functional management identifies and supplies appropriate IT and IPT members for the program.
Training	Program director approves program training needs.	IPT leaders identify special training (program or core discipline) requirements and notify functional management.	Functional management provides for individual contributor development (training, awards, and promotions).
Performance assessment	Program director and IT evaluate program and IPT performance versus program plan.	IPT monitors and reports performance versus program plan.	

and to the major phases of the process framework, from concept design through in-service support.

It is desirable to have direct customer involvement during engineering programs for several reasons. Most importantly, the customer gradually assumes responsibility for the new product over the life of the program. Therefore, the customer should, and in most cases will, jointly participate in the design of the product. Customer personnel will also be better prepared to use the product in a service if they participate in designing it. Moreover, through hands-on involvement with the product as it is designed, the customer's technical staff will be better able to maintain it after it is operational.

During the detailed planning of each phase, a representative for the customer should help determine tasks that should involve customer personnel. Although the workplan templates contain the tasks typically involved in a product development program, they do not completely address all the customer tasks that may be necessary for a particular program. For example, the estimating guidelines for the task do not include time spent by the customer during the reviews, only the time spent by your company.

The customer also helps identify the deliverables in which they are most interested and assists in determining the amount of time required to complete their function on each task. All this information helps the program director build a plan that accounts not only for your company time and responsibilities, but also for the customer's. The detailed plan will enable the customer to budget sufficient time for its efforts and to give you the commitment necessary to make the program a success. The program director will need to guide the customer through the planning process by explaining the planning approach described in Chapter 7.

At the program's outset, the customer should assist in selecting program review board participants and appropriate metrics of design quality. This selection may already have been made during the creation of the program plan in the concept design phase and, therefore, may only need confirmation here. In addition, customer management should participate in briefing the program team on the program's scope and objectives. This involvement shows the customer's commitment to the program and helps motivate the program team.

During the planning for infrastructure changes, team training should be scheduled. Customer/partner management should participate in this effort by helping to determine the appropriate training for their team members (e.g., CAD/CAM and Design for Manufacturing). Just as important as their involvement in planning and initiating the program is the customer's role throughout the program's execution. As part of a joint ownership approach, the customer should work hand in hand with the program director, other company professionals, partners, and major vendors. The customer should actively participate in the following activities:

○ Defining the business objectives and product requirements
○ Evaluating subcontractor proposals

○ Attending subcontractor presentations

○ Assessing new product technology and formulating product concepts

○ Preparing product/component marry-up strategies

○ Planning for product/component marry-up with existing products or installations

○ Reviewing program work products

○ Making decisions affecting maintainability

○ Testing the product

○ Participating in evaluating design quality and program reviews

○ Participating in program replanning points

○ Assisting in the program's overall management and control

○ Assisting in managing engineering changes during the program through the review, approval, and funding of engineering change requests

When the customer actively participates, a wealth of knowledge and experience regarding its needs is brought to the program. Although your company will have considerable industry and technical experience, you should recognize that each customer has a unique set of needs.

Placing customer personnel in key program level roles is important in maintaining an appropriate level of the customer's engineering staff involvement. However, there are certain stages of the program at which their participation is particularly important. As with all recommendations in the guidebook, these guidelines must be tailored according to the particular program and customer characteristics.

The program director's efforts to involve the customer throughout the program should result in accurately defined requirements, timely resolution of issues, a smooth transfer of responsibility, and, ultimately, a business need that is satisfied effectively. In other words, having constantly involved the customer should help ensure that the new product really does benefit the customer. A word of caution is necessary. The customer does not have the same interest in containing the statement of work as your company has. A buttoned-down SOW is critical to managing customer involvement.

4.2 THE PARTNER

Many engineering efforts today combine the efforts of more than one company. Numerous examples abound in the aerospace, automotive, and electronic industries. These relationships take the form of partnerships. The following explanation details the roles and responsibilities of a partner's involvement in a program. A discussion regarding the critical success factors of partner involvement is also included . For a quick reference, Tables 4-1 through 4-3 show the responsibilities in relation to the subphases within program management and to the major phases of the process framework, concept design through in-service support.

The partner's engineering staff plays a key role in a program's success and should participate equally with your company whenever possible. Their understanding of the current product technology, design approaches, and CAD/CAM technology is invaluable, especially if your company has not done much engineering work for the partner in the past. Having a program director and active team members assigned to the program gradually promotes the partner's sense of ownership for the product design. It also gives them the knowledge they will need to integrate their design with your design. One of the key strategic benefits of partnering is technology transfer. You should ensure that your team is prepared to learn as much as possible from the partner's staff. Because they will employ different design rules, you must seek to understand the differences rather than leap to judgments.

At the program's outset, the partner should assist in selecting appropriate metrics of design quality and program review board participants. (This selection may already have been made during planning and, therefore, may only need confirmation here.) In addition, customer/partner management (in particular, the sponsor) should participate in briefing the program team on the program's scope and objectives. This involvement shows the partner's commitment to the program, which helps motivate the program team.

Program initiation is the point at which team training should be scheduled. Partner management should participate in this effort by helping to determine the appropriate training for program team members since they know these members' background and skills. The partner should participate in the following activities:

○ Assisting in monitoring the program's scope, budget, and schedule
○ Evaluating subcontractor proposals
○ Attending subcontractor presentations
○ Assessing new product technology and formulating product concepts
○ Preparing product/component assembly strategies
○ Planning for product/component assembly
○ Reviewing program and subproject work products
○ Testing the product
○ Participating in evaluating design quality and program reviews
○ Participating in program replanning points
○ Assisting in the program's overall management and control
○ Assisting in managing engineering changes during the program through the review, approval, and funding of engineering change requests.

The program director should also insist that the partner's executive management review the program team's work. Your company's corporate personnel understand how the business works; they can help you ensure that all requirements are clearly stated and that operational and policy impacts are identified. Engineering management knows the product integration issues and can provide insight into why the product was conceived in a particular manner. Both your company's corporate management and its engineering management can help you anticipate problems and avoid difficulties throughout the program by evaluating the design as it progresses.

4.3   THE SPONSOR

The roles and responsibilities of a sponsor's involvement in a program is discussed here and is followed by an examination of the critical success factors of sponsor involvement. For a quick reference, see Tables 4-1 through 4-3, which show the responsibilities in relation to the various product development phases.

The program should have a strong sponsor. The program sponsor has ultimate authority over and responsibility for the program. The sponsor may come from corporate or one of the divisions. The sponsor is an executive who has a vested interest in the results of the program, funds the program, resolves conflict over policy or objectives, and provides high-level direction. Because the sponsor can take executive action to resolve issues and conflicts on the program, your success is closely tied to the sponsor.

The sponsor need not have detailed engineering design experience because his or her role is primarily that of a business decision maker. These responsibilities involve linking the day-to-day design activities with the company's strategic direction and financial commitments.

The sponsor is responsible for approving major engineering changes to the product at each phase. During the final stage of the program, the sponsor and the customer's executives are responsible for accepting the product definition as ready for full-scale production. Therefore, the program director should check that customer personnel have completed acceptance testing and are satisfied with the results. At this stage, various customer and partner personnel participate with the sponsor in monitoring and fine-tuning the final product definition. The program director should work with them to coordinate closely the resolution of any outstanding problems or acceptance issues. The process to perform this work should be clearly defined in the documentation of your company's Define process.

4.4   THE PROGRAM DIRECTOR

The following discussion explains the roles and responsibilities of a program director and the critical success factors of program director involvement. For a quick reference, Tables 4-1 through 4-3, show the responsibilities in relation to the various product development phases. Program management requires the involvement and integration of numerous functional organizations. The roles and involvement of other support functions such as finance, contracts, and business

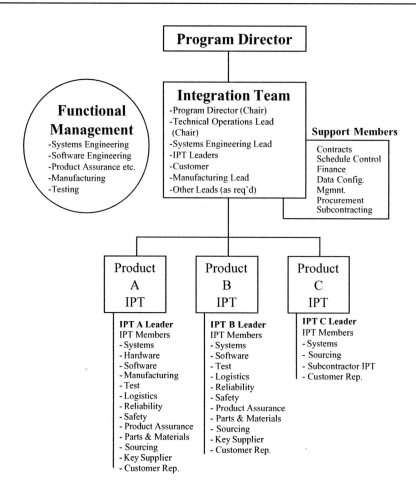

Figure 4-1: Sample program organization

Program Management	Integration Team	Integrated Product Team
Program Director: integration team (sometimes referred to as a program integration team) leader reporting to business area director or company president as appropriate.	Integration team: multifunctional team including support organizations and lower tier IPT leaders (sometimes implemented as a program integration team, and systems engineering and integration team). Acts as proposal team during proposal effort	Integrated product teams: multifunctional team including support organizations. IPT leader: Assigned by IT and functional management and is responsible for team coordination and reporting to IT. Acts as cost account manager. Acts as estimating team during proposal effort

development are not addressed in this book because of their high degree of dependence upon your company's unique business processes. Figure 4-1 illustrates a sample program organization.

The program director is responsible for the overall technical, schedule, and cost performance for the program, including functional cost management.

Additionally, the program director assumes the following responsibilities:

- ○ Serves as the primary customer interface and the integration team chair
- ○ Ensures that methodologies are implemented and followed on the program
- ○ Ensures compliance with applicable policies and procedures
- ○ Directs the program's planning, estimating, and scheduling activities and risk plan
- ○ Approves program training needs
- ○ Approves resource requirements and manpower phasing
- ○ Identifies the required IPTs and dissolves them as required
- ○ Selects integration team members (with functional organization concurrence)
- ○ Provides progress reporting structure and reports
- ○ Takes leadership of the program
- ○ Takes technical responsibility within partners
- ○ Takes responsibility for engineering cost, quality, and schedule
- ○ Coordinates efforts between customer, partner, subcontractor and other personnel
- ○ Participates in presentations to the customer or partner

Successful programs are the result of an organized, carefully thought out and intelligently applied, motivated team effort. That effort is largely up to the program director who must see to it that personnel at all levels of the company understand the program. The program director also must maintain a broad awareness of the status of major activities and a detailed knowledge of any problems that arise. Finally, the program director must look after and coordinate the particular needs and efforts of the customer, partner, and company's team members.

One of the most challenging tasks as a program director is to maintain an appropriate level of customer involvement throughout a program. This is especially critical on multipartner programs, where efforts to work with the customer distinguish you from the competition. Many other companies fail to work with the customer to solve the business problems.

The program director works closely with the partner's program director in addressing the needs of the customer and in coordinating joint production efforts.

As the detailed plan is developed, the following categories of customer personnel should be considered for involvement:

- ○ Top corporate management
- ○ Users of the existing products (all levels)
- ○ Engineering staff
- ○ Customer support personnel

A program director should work with the customer's engineering staff to understand the customer's needs thoroughly before proceeding to design a product.

Using techniques such as 3-D modeling or virtual mockups will help stimulate the customer to identify requirements. The experience and input of the customer's staff can lead to an earlier and better understanding of problems and constraints.

Regardless of the level of customer involvement, the program director will benefit from a positive working relationship with all customer personnel. As part of this effort, regular progress meetings should be held with customer personnel. Recommended progress reporting procedures are outlined in Chapter 9. The program director has primary management responsibility for the entire program, including administration, planning and scheduling, issue resolution, customer/partner/corporate reporting, and, with the functional directors, technical leadership.

In the matrix environment used to manage most engineering programs, functional management usually controls program resources. Therefore, it is important that the program director set clear plans and goals for the program to create buy-in and team spirit among the program members.

In day-to-day work, team members can feel torn by seemingly conflicting priorities. The program director and senior engineering functional management are responsible for clearly identifying the relative priority of programs.

The program director's ability to direct a program depends largely on how well he or she applies managerial strengths, both human and technical. Human managerial strengths involve leading, guiding, and inspiring your team. It requires highly developed interpersonal skills such as negotiation, persuasion, and the ability to listen. On the technical side, the program director must be able to plan, organize, measure, evaluate and control performance, assess risk, estimate, schedule, and conduct performance evaluations. These strengths, together with the appropriate techniques and tools, are the keys to effective program management.

In terms of qualifications, the program director should have the company's unique industry experience, a working knowledge of the engineering management system, and experience in the technical disciplines and product areas involved in the program. In addition to professional engineering qualifications, it is important to also have postgraduate business qualifications such as an MBA or similar commercial training.

4.5 THE FUNCTIONAL DIRECTOR

The roles, responsibilities, and critical success factors of the functional director's involvement in a program are examined in this section. For a quick reference, Tables 4-1 through 4-3 show the responsibilities in relation to the various product development phases. Program and product technical performance requires the involvement and integration of numerous functional specialists and organizations.

This involvement begins when the program manager and integration team contact the functional director.

Functional management in an IPT environment supports the program director and integration team by providing the program with competent, trained personnel from their organization. In addition, they

- ○ Provide and maintain functional processes, policies, and procedures
- ○ Provide technical expertise and consultation at program reviews (including cost and risk expertise)
- ○ Provide equipment and facilities required by IPTs
- ○ Facilitate training needs, and manage the career progression of the IPT members
- ○ Maintain core technical competencies within their organization
- ○ Ensure that functional processes and procedures are followed by the IPTs

In addition to technical expertise functional directors need to have command of their organization's resources, capabilities, and processes. They must ensure that their functional processes are integrated into the programs. These directors should also possess a thorough knowledge of the new product development processes.

In general, involvement of functional management serves to ensure that all technical, cost, schedule, and quality requirements for the product are met.

During concept design and upfront planning, the functional director is asked to contribute to the compilation of a program plan. The deliverables during this phase include

- ○ A list of tasks to be completed
- ○ The logical precedents of tasks required to be completed before the task can begin
- ○ The estimate of hours required to complete each task
- ○ The skills or resource pools from which skills will be drawn to complete the task
- ○ Technology to be applied to the business processes

Figure 4-2 provides an example of a functional organization in an aerospace company. In addition, the functional director must be able to manage competing egos within the program. Often program directors/managers look down on functional management, describing them as "overhead," and employees frequently see this function as a place where the least capable are placed. This assumption is incorrect; the functional director must manage these views carefully. See Chapter 13 for techniques required to do this.

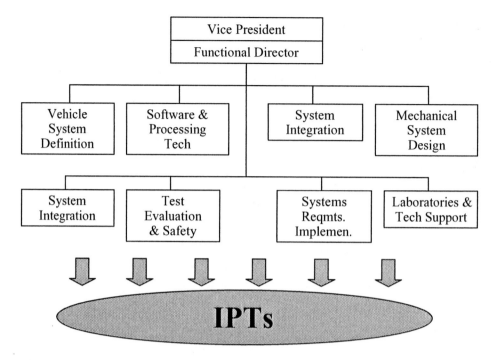

Figure 4-2: Engineering functional organization

4.6 THE PROJECT MANAGER/ENGINEER

This section examines the roles, responsibilities, and critical success factors of the project manager/engineer. For a quick reference, Tables 4-1 through 4-3 show the responsibilities in relation to the various product development phases.

The program manager/engineer is responsible for clearing any barriers that might get in the way of the functional team members as they try to complete the task. The project manager/engineer typically leads an IPT. As such, the project manager/engineer leads an IPT to translate the marketing requirements and objectives into design requirements. Later that is translated into a design spec.

The first step in planning involves the high-level structuring of the program. This effort involves tailoring the objectives and deliverables of each subphase to meet the specific needs of the program, reviewing the recommended program review points, and determining the ones appropriate to the IPT project. It also includes establishing IPT project procedures (e.g., engineering change control and program reporting). These reviews coincide with the phased structure of the define process architecture. Involving customer/partner personnel in creating the program plan (prepare for management review and approval) will ensure that they have a good understanding of the work to be completed, the deliverables to be produced, and the manner in which the program will be conducted. Throughout the program-structuring process, the program manager/engineer and the customer/partner should discuss the program characteristics and expectations. You should reach a consensus on the following items:

- ○ Subphase objectives
- ○ Customer deliverables
- ○ Compliance with design requirements
- ○ Program review points
- ○ Program approach, including
    - ⇨ The concurrent product development approach
    - ⇨ The program management approach
    - ⇨ The business process framework

The responsibilities of a program manager/engineer include

- ○ Planning the work that will realize the desired objectives
- ○ Structuring the WBS and determining the level of planning and reporting
- ○ Scheduling/organizing the work to meet the objectives, through the program coordinator
- ○ Working with functional management to staff the program to accomplish the work
- ○ Motivating the staff to do the work so that the objectives will be achieved
- ○ Controlling the work by ensuring that corrective actions are taken to help achieve the objectives

The most important job of the program manager/engineer, however, is to receive the estimate prepared by the functional managers and then create the program's process workflow diagram. The program manager/engineer then creates an IPT program schedule that meets the following criteria:

- ○ Respects the demanded completion date
- ○ Respects the limits of resources available in the resource pool

If conflicts for resources from specific pools arise between programs, the IPT project manager/engineer must resolve the conflict using one of the following three avenues:

- ○ Adjust the program schedule without impacting the committed completion date
- ○ Adjust another program schedule without impacting its committed completion date
- ○ Identify the problem for the program director who will reprioritize the programs.

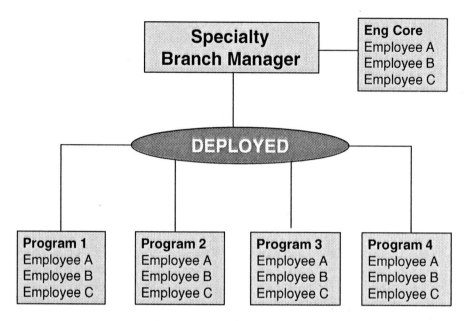

Figure 4-3: Functional deployment of engineers

The planning process also entails a risk assessment so that risks can be effectively managed throughout the program (see Chapter 8). The customer and/or partner can assist in this effort, and they are generally eager to do so because they have a vested interest in minimizing the program's chances of failure. They can provide valuable insight into the program's risks because they understand the market needs and, in some cases, the product integration requirements. In addition, customer involvement in risk assessment helps establish the initial bonds that lead to a good working relationship.

4.7 THE FUNCTIONAL MANAGER

The roles, responsibilities, and critical success factors of the functional manager are described in this section. For a quick reference, Tables 4-1 through 4-3 show the responsibilities in relation to the various product development phases. Figure 4-3 illustrates how a functional manager may deploy various engineers to many programs. Often an engineer may be involved in two or more programs at the same time.

After the program is under way, the functional manager is required to

○ Manage the resources within each pool to ensure that the committed work is done on time
○ Report actual performance against the tasks committed to be done and show hours remaining for any task in progress

Some of the responsibilities that the functional manager should provide are

○ Skills (design, analysis, tooling, methods, testing, etc.)
○ Technology to be adapted on a business process
○ Training requirements of IPT members
○ Understanding of workflow (business process, composition, and deliverables architecture)
○ Monitor performance of tasks being completed to schedule (leadership)
○ Capacity planning

The key roles of functional management are to:

○ Provide competent, trained personnel
○ Provide "technical" expertise and consultation
○ Maintain core technical competencies
○ Develop and maintain core systems/processes
○ Provide equipment and facilities
○ Accomplish infrastructure and "common" tasks
○ Establish metrics and benchmarking
○ Manage the workforce (hiring, career and succession planning, training, promoting, ranking discipline, and downsizing)

4.8   THE PROGRAM COORDINATOR

The following explanation details the roles, responsibilities, and critical success factors of the program coordinator's involvement in a program. For a quick reference, Tables 4-1 through 4-3 show the responsibilities in relation to the various product development phases.

The program coordinator who often leads an integration team helps to avoid "stove pipes" among the IPTs and ensures that horizontal exchange of information occurs among program IPTs and ITs. Their roles and responsibilities on a program include

○ Integrating the overall product/system
○ Allocating resource and cost targets for the IPTs
○ Establishing the program product structure and WBS
○ Organizing the IPT
○ Developing the integrated master plan and integrated master schedule
○ Gaining functional concurrence on technical plan, risk plan, staffing, and manpower phasing
○ Developing colocation and enabling technologies plans for the program
○ Coordinating a program IPT trade study process
○ Monitoring overall technical, cost, and schedule performance of the IPTs

Some of the responsibilities that the program coordinator should provide are

○ Program management
○ Customer interface
○ Overall program performance
○ Estimates, budgets, costs, and schedules
○ Management of IPTs – creating and dissolving them
○ Resource requirements
○ Infrastructure tasking and "common" support
○ Day-to-day IPT personnel tasking including
    ⇨ Annual objectives
    ⇨ Performance
    ⇨ Merit awards

The program coordinator's ability to direct a program depends largely on how well he or she applies managerial strengths, both human and technical. Human managerial strengths include leading, guiding, and inspiring the team as well as basic interpersonal skills such as negotiation, persuasion, and the ability to listen. On the technical side, the program coordinator must be able to plan, organize, measure, evaluate, and control performance, all of which require the program coordinator to be adept at assessing risk, estimating, scheduling, performing evaluations, and forecasting.

These strengths, together with the appropriate techniques and tools, are the program coordinator's keys to effective program management. Figure 4-4 illustrates the relationship between program management and functional management.

The program coordinator should have industry experience, a strong knowledge of the methodologies described in this book, and experience in the technical disciplines and product areas involved in the program. In addition to professional engineering qualifications, the program coordinator should round out his or her training with postgraduate business qualifications such as an MBA or similar commercial training.

4.9 THE IPT LEADER

In this section, the roles, responsibilities, and critical success factors of the IPT leader involvement in a program are explained. For a quick reference, Tables 4-1 through 4-3 show the responsibilities in relation to the various product development phases. Figure 4-5 illustrates the need for the IPT leader to be at the center of the team in a participatory style.

The IPT leader is responsible for the day-to-day management of a subset of the program. That subset might be a particular subsystem (e.g., wings or gear train), a technical area (e.g., stress analysis or tool design), or another division of

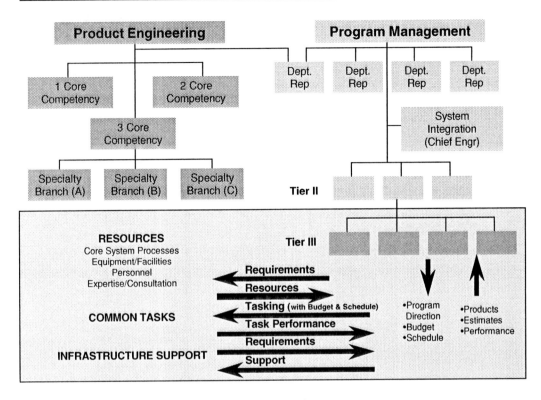

Figure 4-4: Program management implementation format

Team Leader / Chairman

Participatory, Self-Directed Team Management

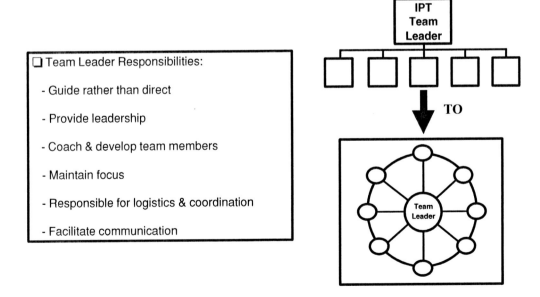

❑ Team Leader Responsibilities:

- Guide rather than direct

- Provide leadership

- Coach & develop team members

- Maintain focus

- Responsible for logistics & coordination

- Facilitate communication

Figure 4-5: IPT team leader responsibilities

the program, based on the particular problem being addressed. The IPT leader's role includes

○ Administrative reporting to the program director
○ Supervising team members
○ Assisting in overall program planning
○ Providing scheduling and technical leadership
○ Advancing team building and group dynamics

The integrated product team leader is a team member with added responsibility for supervising tasks, providing resource management, and nurturing the team environment. The team leader's authority is most evident when running meetings and leading discussions. Selection of a team leader with direct supervisory authority should be carefully considered because there is a temptation to allocate responsibility to the team leader rather than to the entire team, with a resulting loss in team synergy. Clearly, IPT leaders should possess the strong leadership skills necessary to foster the IPT environment.

The IPT leader is accountable to the integration team for the content and progress of IPT activities. He or she ensures that progress is made toward IPT goals and objectives including technical approach, budget, schedule, and quality and that relevant information is not omitted. The IPT leader is also responsible for ensuring that all required IPT roles are filled, that IPT meetings are efficient, and that group documentation is developed and made available. The leader monitors performance data and initiates corrective actions to ensure that the IPT functions within its budget and schedule, that risks are mitigated, and that technical requirements are met. In this sense, the IPT leader also acts as the cost account manager for the IPT's product deliverable. The team leader also ensures that there is coordination up, down, and across the program structure. The leader is the initial point of contact for other IPTs and sponsors, but he or she is not solely responsible for the product deliverable that is the joint responsibility of the IPT and ultimately the program director. The team leader also resolves issues of competing needs within the IPT.

Specific IPT leader responsibilities include:

○ Lead the team in coordinating technical activities
○ Lead the team in carrying out the statement of work, with the overall responsibilities of meeting the integrated schedule, estimate, and quality requirements
○ Lead the team in resolving all issues that have a potential effect on the team's performance
○ Identify for the program coordinator any issues that cannot be resolved at the team level, including a full description of the issues precluding a

decision being made at the team level and a demand date to the program coordinator for a decision that will support the team's schedule
○ Involve the appropriate functional representatives in team decisions, as and when required
○ Ensure that the team works as a unit dissolving barriers between the functional organizations
○ Keep both the program coordinator and the IPT informed of progress and potential problems
○ Ensure daily communication with team members and hold formal meetings as required
○ Monitor compliance with the integrated schedule

The IPT leader's ability to perform effectively depends on the ability to interface with the other IPT leaders, program coordinator, and functional managers. In addition, the IPT leader will require very dynamic skills to balance both the technical and emotional needs of team members.

4.10 THE TEAM MEMBER

The roles, responsibilities, and critical success factors of the team member's involvement in the program are examined in this section. For a quick reference, Tables 4-1 through 4-3 show the responsibilities in relation to the various product development phases.

Figure 4-6 illustrates the changing roles of an IPT from phase to phase. The team member will receive various assignments from the program and functional directors.

Understanding how the engineering department organizes information will permit a team member to find and assimilate requirements quickly to complete the assignment. The approach to completing an assignment is described in Chapter 6.

The primary responsibility of IPT members is to fully understand the scope and impact of technical, budget, and schedule issues confronting the team and to communicate the impact those issues have with respect to their particular area of expertise to the other team members and IPTs. This communication forms the fundamental process for resolving problems during the preliminary design stage of a program.

IPT members must have the authority to contribute equally and to commit their departments to team decisions. In addition to representing the interests of the functional organization, each IPT member must also support the IPT plan. Many roles are available on an IPT, such as serving as a point of contact with other IPTs or the IT, recording meeting proceedings, or leading team discussions and trade-off studies. To enable the IPT to function openly and candidly, each

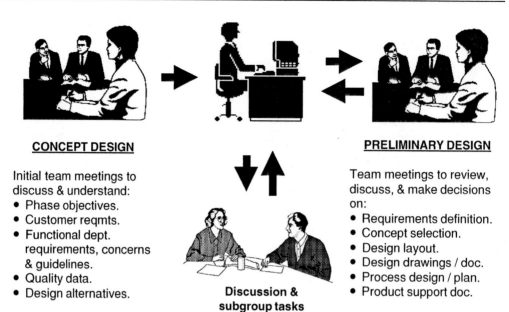

CONCEPT DESIGN

Initial team meetings to
discuss & understand:
- Phase objectives.
- Customer reqmts.
- Functional dept.
  requirements, concerns
  & guidelines.
- Quality data.
- Design alternatives.

**Discussion &
subgroup tasks**

PRELIMINARY DESIGN

Team meetings to review,
discuss, & make decisions
on:
- Requirements definition.
- Concept selection.
- Design layout.
- Design drawings / doc.
- Process design / plan.
- Product support doc.

Figure 4-6: Integrated product team operation deliverables by phases

team member must be both willing and prepared to fill any of these roles when necessary. Rotation of specific responsibilities among the team members is a way to prepare and encourage members to participate proactively.

Another IPT member responsibility, which is critical in the IPT environment, is to resolve problems. Traditionally, a specific individual or team has been assigned responsibility for resolving problems. In the IPT environment, this responsibility belongs to the entire IPT, not to a specific individual or specialized team. The IPT takes responsibility for the success, failure, or problems associated with their product development plan. IPT members must understand this concept and accept this responsibility. With empowerment comes full accountability.

In a traditional philosophy of product development, the person designated as team leader is expected to communicate with management and other teams. In an IPT environment, the team, not an individual, is responsible for team communication. The team leader requests information, but any team member may provide the information. Instead of expecting one person to know everything about a program, all team members are forced to keep aware of team developments, and the individual best equipped should provide the needed information. Members of a tightly working team cross-train each other. Figure 4-7 illustrates the responsibilities of the IPT.

To effectively commit their support to team decisions, team members need a strong working knowledge of their functional organization's resources and processes and how they relate to the business, in addition to technical expertise in their disciplines.

- Empowered, self-directed, ownership - team is responsible.

- Balanced, consensus decision making.

- Establish and maintain workplan.

- Establish team ground rules.

- Working sessions and formal team meetings - action items and minutes.

- Reviews and approves drawings, other design documents, etc.

- Process improvement.

- Status reporting.

Cross-Functional Web

Figure 4-7: Team member responsibilities

4.11 SUPPORT FUNCTIONS

This section describes the roles, responsibilities, and critical success factors of support functions involvement in a program. For a quick reference, Tables 4-1 through 4-3 show the responsibilities in relation to the various product development phases.

Most programs require additional personnel to support the program's administration. These support functions should not be overlooked during planning and should be provided for in the program organization. The amount of resources needed to support a program will vary based on its size. The responsibility for support functions may be spread across many administrative staff. Some of the most important of these support functions, grouped by type, are presented here.

Administrative – Business management

Organizing the program office – Ordering program supplies, equipment, engineering process documentation and automated tools; assigning space and equipment; establishing the program library and program repository.

Coordinating services to the program team – Coordinating telephone and reception services; setting up and coordinating word processing activities; coordinating mail and courier services.

Coordinating assembly and reproduction of customer deliverables – Arranging graphic support; coordinating the assembly and reproduction of materials.

Reporting

Supporting time and expense reporting – Collecting and validating program time and expense sheets; entering time and expense information into the program management system; reconciling time and expense charges to the accounting system.

Supporting program status reporting – Assembling program status information from the program team; generating draft status reports for the program director; generating financial program status reports.

Supporting billing – Generating customer billing documents; generating billing invoices and memoranda.

Monitoring subcontractor activity and expenses – Validating vendor invoices; reviewing subcontractor fees and living expenses; filing contracts; generating time and expense reports.

Scheduling and coordination of IPT activities

Coordinating continuing education – Enrolling the program team in courses on the management bodies of knowledge and the holistic approach, DFMA, automated tools, or program-related topics.

Coordinating schedules – Coordinating program team vacation schedules; communicating the location of team members; maintaining the automated program schedule and workplan.

Critical success factors

The support functions must be highly responsive to the IPTs. Organizational bureaucracy must be kept to a minimum. Streamlined business processes between support functions and the program will ensure effective responsiveness. Often engineers look down on these roles as "clerical" and carry this attitude. Without the key support roles, all programs would be in chaos; therefore, respect for support roles is essential.

4.12 SKILL TYPES

We recognize that different types of skills are required to complete an engineering program. For each workstep, a skill type has been identified and is included on the workplan templates used in assigning work to individual team members. Skill types do not necessarily equate to job titles. Rather, they are roles that need to be filled on a program and that require specific skills. In a small program, one person may fill several roles (e.g., manufacturing engineer, tool design, NC programming, methods), whereas the same role may be filled by several people in a large program.

Some worksteps, such as performing a walkthrough, may require more than one skill type. The workplan templates indicate this by citing "program IPT" as the appropriate skill type.

After team activities have been geared to "milestones within phases" and "tasks within milestones," it is now possible to measure the process with a

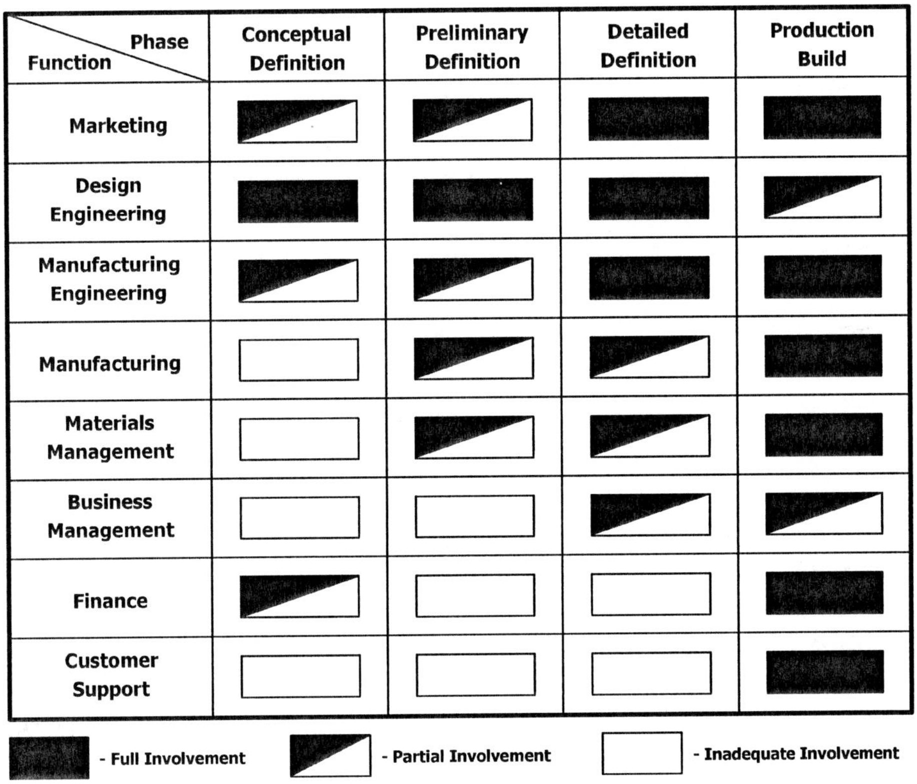

Phase / Function	Conceptual Definition	Preliminary Definition	Detailed Definition	Production Build
Marketing	Partial	Partial	Full	Full
Design Engineering	Full	Full	Full	Partial
Manufacturing Engineering	Partial	Partial	Full	Full
Manufacturing	Inadequate	Partial	Partial	Full
Materials Management	Inadequate	Partial	Partial	Full
Business Management	Inadequate	Inadequate	Partial	Partial
Finance	Partial	Inadequate	Inadequate	Full
Customer Support	Inadequate	Inadequate	Inadequate	Full

- Full Involvement   - Partial Involvement   - Inadequate Involvement

Figure 4-8: Concurrency matrix

concurrency matrix, which is illustrated in Figure 4-8. For example, manufacturing engineering has partial involvement at the concept and preliminary definition phases and full involvement at the detailed definition phases. The concurrency matrix can be used to measure functions involved across the life cycle. A judgment can then be made to determine if the effort is balanced.

4.13 TRAINING GUIDELINES

To provide high-quality service in solving the customer's business problems, the IPT must be trained in the management and technical skills necessary to fulfill their program responsibilities. Some of the technical skills for which training may be needed include the use of the process integration framework and automated tools that are to be used on the program. Many courses in engineering specialties are available through the company's professional development program.

The risk assessment completed at the beginning of a program indicates areas in which specific technical training may be needed in order to avoid risk. See Chapter 8 for more information.

Approach to Program and Project Management

An advanced engineering program requires the efforts of many professionals. Although no single contribution is ever responsible for a program's success, a project cannot hope to be successful unless it is managed well, unless all the personnel and resources involved are coordinated and directed toward a common goal.

In any organization, having people play their roles and take defined responsibilities plays a critical part in ensuring success. For example, it is very important that a functional manager take responsibility for running the organization and controlling the utilization of the resources.

This chapter focuses on the processes of program management. Reflecting on the program management workplan template, these pages discuss the major responsibilities that the program and functional directors have throughout the program. Also included is information on the interrelationship of roles.

The objective of this chapter is to outline the approach to managing a project. We stress that planning and control is an important component of any program management technique, but we will also look at specific leadership issues facing the program and functional directors, the importance of creating executive commitment to your program, and the interrelationships of key roles.

5.1 WHAT IS PROGRAM/PROJECT MANAGEMENT?

The complexity of program/project management makes it difficult to describe in a few words. Basically, however, it can be defined as planning, organizing, directing, and controlling resources to meet a specific business goal. In an advanced engineering program, sound program/project management means fulfilling the program's objectives on time and within budget, making effective use of the assigned resources, maintaining good customer and partner relations, and providing a product that meets the customer's needs.

Within the context of this book, the program/project management process provides three main tools for helping the functional or program director conduct his or her work. This process must respect the concept of roles and responsibilities explained in more detail in Chapter 4. The tools are

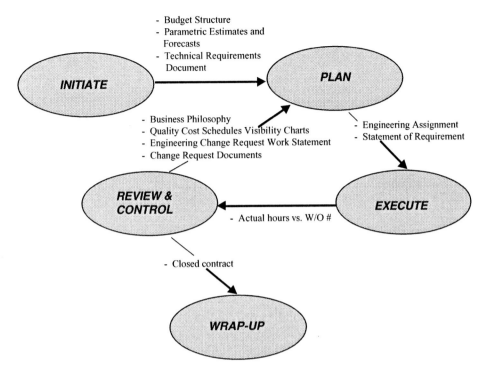

- Budget Structure
- Parametric Estimates and Forecasts
- Technical Requirements Document

- Business Philosophy
- Quality Cost Schedules Visibility Charts
- Engineering Change Request Work Statement
- Change Request Documents

- Engineering Assignment
- Statement of Requirement

- Actual hours vs. W/O #

- Closed contract

INITIATE
PLAN
REVIEW & CONTROL
EXECUTE
WRAP-UP

Figure 5-1: The five major processes of project/program management

○ Project planning templates
○ Engineering reference manuals
○ Resource management techniques

The program phases of the process framework provides for five major processes of program management (see Figure 5-1).

○ *Initiating*
This process includes such activities as authorizing the program and preparing the budget structure, parametric estimates, and forecasts and technical requirements.

○ *Planning*
This process includes such activities as releasing work, providing guidance, and coordinating the efforts of the team. It requires that issues uncovered during the initiation be addressed and also focuses on the development of structure for the project.

○ *Executing*
This process includes such activities as determining status and evaluating quality and productivity. Establishes the means for monitoring, reporting, and taking corrective action for the project, through the process framework.

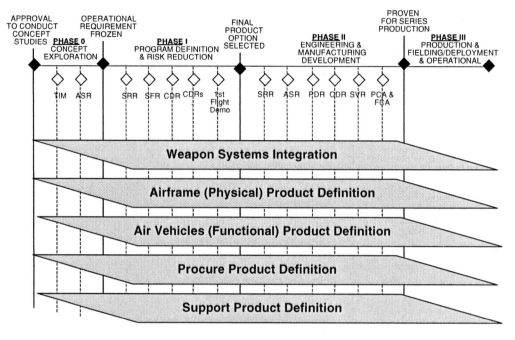

Figure 5-2: Sample resource load by function and phase

This is the primary process for carrying out the project plan. The program team must manage the various technical and organizational interfaces that exist in the project.

○ *Review and controlling*

This process monitors the program plan in relation to the established cost, quality, and schedule objectives. It ensures that the objectives are met by measuring progress and taking corrective actions when necessary, throughout the process framework.

○ *Wrap-up*

This process includes such activities as reporting resource expenditures and project status to the customer and corporate management. It also includes ensuring that the data repository and lessons learned are properly documented. This process also formalizes the acceptance of the program and brings it to an orderly end.

These program management processes are used to manage each phase of the program. In our example in Figure 2-1, we illustrated phases D1 through D7. At the final review of each phase, we review the deliverables of that phase and the plan for the next phase. The next section of this chapter discusses the overall responsibilities in meeting that challenge.

Harmonized schedules and properly balanced resource loading are invaluable tools for reducing waste in a program and reducing time to market. A resource

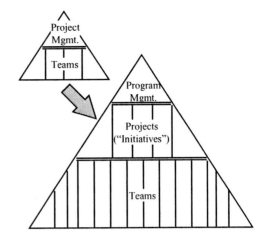

Environmental Factors

- Projects are becoming larger and require multiple skills delivery

- Functions from various parts of the organization must work together

- Due to these developments, the complexity of the projects are increasing

- Hence the task of project management is getting more complex

Project Manager

- Allocates resources
- Tracks progress against schedule
- Checks cost against budget

Program Manager

- Manages resources for several projects at the same time showing leadership
- Tracks progress of complex projects with many cross-functional teams
- Drives for total benefits within total costs
- Risk manages the delivery
- Produces all reports needed to fulfill the contractual requirements
- Sets up the infrastructure for large teams

Old tools

- Schedule
- Budget
- WIP

New tools

- Program Office
- Cost/benefit model
- Progress tracking methodology
- Risk Management

Figure 5-3: Contrasting project and program management

loading diagram plots over time the capacity for each resource vs. the number of units of that resource required by all committed projects. These diagrams are important considerations when the organization is considering the impact of large new programs. Figure 5-2 illustrates a resource load profile.

5.2   WHY ARE PROGRAM OFFICES NECESSARY?

A program office is particularly useful when the program is decomposed into subprojects. A list of reasons for a program office follow. Figure 5-3 illustrates the relationship of program management and project management. Programs are made up of a collection of subprojects, and program management includes risk assessment.

5.3   CRITICAL SUCCESS FACTORS – SECRETS OF SUCCESS

Countless factors can contribute to the success or failure of an advanced engineering program, but the success of any project depends on its *critical success factors*. Before focusing on these, however, it is important to clearly define success. Any program that is completed on time, within budget, at or above the desired level of performance and quality, and within the mutually agreed upon scope of changes and that also satisfies the customer's business objectives qualifies as a successful program.

Different results constitute successes for different projects, but over time we have learned that certain factors must be present if a project is to be successful. Corporate management, the customer, the vice president of engineering, and the program and functional directors all play important roles in helping to ensure that these critical success factors are present in your project. Brief discussions of each of these factors follow.

5.3.1   A genuine understanding of the program

It is essential that corporate management, the customer, the vice president of engineering, and the program director

- ○ Clearly define the expectations for the program with a formal marketing requirements and objectives document
- ○ Clearly define and understand the program's risks using the risk assessment process
- ○ Understand the engineering processes as they apply to the program in a cross-functional way

5.3.2   Executive commitment to the program

Executive commitment includes corporate management, partners, and customers. They should

- ○ Establish reasonable, specific goals that relate to corporate objectives
- ○ Provide the necessary resources
- ○ Show visible support for the program
- ○ Provide for prompt, accurate communications through clearly defined channels
- ○ Take action on requests

○ Facilitate the program team's interface with support departments

○ Provide protection from political in-fighting and manage the political climate

In addition, corporate executives should be careful to provide team members with opportunities for personal and professional growth.

5.3.3   Effective leadership by the program and functional directors

The success of a program depends greatly on the efforts of the program and functional directors. After all, they deal directly with the company, the partner executives, and the customer. They also directly oversee the work of the integrated product team members. Their leadership is crucial to the program's success. This effort includes the following major responsibilities:

○ Managing the expectations of the corporate executives, the customer, and the team members

○ Maintaining a good relationship with the partners and the customer executives

○ Presenting the corporate executives and the customer with recommendations, not just alternatives

○ Providing appropriate resources to the program

○ Ensuring that IPTs practice effective decision-making and problem-solving techniques

○ Tying together responsibility, performance, and rewards when managing team members

○ Minimizing outside pressure on the IPTs

5.3.4   Organizational adaptability – The toughest challenge

During a program, the corporate management, the partners, and the customer need to adapt to one another. Therefore, we should be sensitive to the customer's way of doing business (its regular reporting procedures, channels of communication, etc.). At the same time, the time that is taken to explain to the customer the way the company manages engineering programs will pay off many times over.

Also, a new program will probably necessitate changes in a partner's and the company's organization culture. Here are some ways the program director can help a partner/customer manage these changes:

○ Be sensitive to the organizational and cultural changes the new program requires

○ Identify criteria by which the success of implementing these changes can be judged

○ Identify the factors most likely to facilitate or hinder implementation
○ Plan ahead to overcome resistance

5.3.5   Commitment to planning and control

One of the keys to effective program management is seeing that the right work gets done the right way, the first time. The best way to help ensure this is to plan and control the project. To this end, the program director should work with corporate executives and partners to

○ Develop realistic cost, schedule, and performance estimates and goals
○ Identify risk management strategies in anticipation of potential problems
○ Agree on a process for managing changes, including the appropriate authorizations
○ Provide feedback on program and process performance

5.4   PROGRAM MANAGEMENT ELEMENTS

In this section, we outline the various processes of program management. They include:

○ *Define contractual measures*
   Defines the performance measures that will be used to judge the successful fulfillment of the program.
○ *Establish baseline*
   Uses the defined contractual measures and determines a baseline business case at the beginning of the program. It also defines the starting point and identifies the gaps in cost, schedule, or technology to be closed.
○ *Develop the business case and budgets*
   Develops a business case for each project. This includes cost–benefit analysis and defines the achievement milestones; the resulting program budgets are committed.
○ *Develop and maintain WBS and milestones*
   Defines the work breakdown structures for IPTs. The work is scheduled on a level sufficient to set the major program milestones.
○ *Plan and set up program infrastructure*
   Sets up the necessary office space, copiers, computer networks, telephones, meeting rooms, and administrative support.
○ *Develop and maintain communications plan*
   Develops a communications plan with all the stakeholders (executives, functional directors, manufacturing, partners, suppliers). Coordinates closely with the change management/deployment team.

○ *Staffing and staff succession plans*

Identifies the resources needed for the program, chooses the right people to join the initiative, and plans staff succession.

○ *Collect cost and report*

Installs and maintains the function of processing all time and expense claims from the program.

○ *Risk management*

Uses a documented methodology to identify issues that might be a risk to the success of the program (see Chapter 8). This process helps to assess these risks and set action plans to mitigate them. It also implements and tracks actions and ensures that priorities are established logically.

○ *Program quality reviews*

Plans and coordinates all activities needed to maintain a high level of quality in the work conducted by the IPTs.

○ *Commitment planning and management*

Ensures that the benefits and particular resource profiles of the concurrent process are integrated into the budget. The financial support organizations are responsible for validating benefits.

○ *Deployment and rollout*

Coordinates with the team responsible for process management to ensure that tools and processes are properly implemented. This process also manages the concurrence and the dependencies between the various initiatives and minimizes the level of disruption in the organization when more than one improvement initiative is implemented at the same time.

○ *Progress monitoring and cost management*

Uses the earned value principle to check the progress of program execution vs. the plan.

○ *Control benefit streams vs. realization*

Tracks the actual benefit streams vs. the master plan and provides executives with the assurance that the requirements from the business plan have been fulfilled.

○ *Program change control*

Controls the progress to assess the need to make changes to the original plans. This includes changes in the organization, scope, and processes. If a need to make a change is identified, then the process to revisit the original plans starts.

5.5   PROGRAM MANAGEMENT AND BUSINESS INITIATIVES

On large company-wide change initiatives such as implementing engineering process management or ERP (enterprise resource planning), many smaller projects will be executed concurrently. These are called *business initiatives*, and they often happen simultaneously with a new product development program. They will cover improvements in areas such as product development, strategic sourcing, operations management, and process reengineering.

The changes inherent in the initiatives will often depend on each other. This makes it necessary to install some form of leadership over the entire set of business initiatives. In this book, we describe the work to be performed by those leading changes. The principles will apply equally to a program office at the division or corporate level set up to manage ERP, a supply chain improvement, or other business initiatives.

5.5.1   Program management concepts and principles

Program management of large broadly based business initiatives, such as implementing engineering process management, extends beyond the familiar project management concepts in several ways.

○ It is a formalized management process requiring a formal "program office" for its execution.
○ It is integrated with the company's formal management structure.
○ It is both a *leadership* function and a *management* function.

Why is a distinction being made between leadership and management? Most people tend to think of program management in terms of its more visible management elements including budgeting, cost and schedule performance measurement, and benefits tracking. But program management in this context is also proactive. Successful program management provides the vehicle for *communication* within the project and across the extended enterprise. It is the "glue" that holds together a myriad of subprojects, process improvement teams, and information systems initiatives. It also communicates upward to company executives. Moreover, it transcends conventional projects in its scope. It is a partnership of your company, software suppliers, partners, and management consultants, making little, if any, distinction among corporate affiliations. In this environment, financial and performance risks are often shared among these parties. This takes leadership as well as management.

5.5.2   The "soft" side of program management is leadership

If the "hard" side of program management is planning, tracking, and reporting; the "soft" side is leadership. Program management embraces communications and creativity and works at overcoming organizational and political obstacles. It is a process of sharing insights, managing risks, and celebrating advances. It is a high-profile position and balances the commercial needs of the business with technical program execution and the human needs of team members.

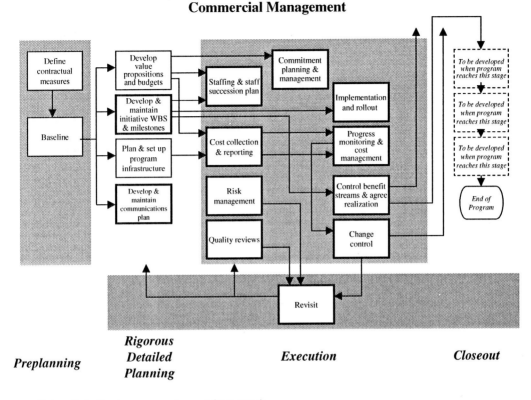

Figure 5-4: Program management framework

Program management and change management go hand in hand. Program management ensures clarity of the program's goals, commitments, progress, and results. The focus of organizational change management is to create understanding and commitment at all levels of the enterprise to the goals. It aims to equip this broad group to act effectively in the program's implementation. It is the foundation upon which day-to-day management is based.

5.6 FRAMEWORK FOR PROGRAM MANAGEMENT

The program modules are shown in Figure 5-4. Each program module is a specific responsibility of the program office and should be discussed with the program director when he or she is appointed. Those modules, represented by boxes with thick frames, are ongoing activities that continue to operate throughout the program. It is important to note that program management is not a process that closely follows one path, rather it is the framework that shows some logical links between the different modules that comprise program management.

5.7   ELEMENTS OF THE PROGRAM OFFICE

The program office manages the following administrative elements:

- ○ Tracking costs and benefits
  - ⇨ A model is used to define benefits and the related cost.
  - ⇨ The model builds on a work breakdown structure defined by the teams from the different subinitiatives.
  - ⇨ Based on this model, tracking is done across numerous projects operating in parallel.
- ○ Creating the necessary reports to fulfill contractual requirements
  - ⇨ Partners in the contract have special requirements with respect to reporting.
  - ⇨ The program officer collects the necessary data and creates and distributes the reports.
  - ⇨ Customized reporting is especially needed in benefit-sharing agreements.
- ○ Processing of bills related to the program as well as all time and expense reports from the staff and consultants on the initiative
  - ⇨ To allow prompt cost tracking, accounting is done on site.
  - ⇨ For staff time and expenses, a harmonized form of data gathering is required to yield correct cost information.
  - ⇨ In larger assignments, staff from a variety of company divisions with differing procedures and policies are involved.
  - ⇨ The program office deals with all these procedures and policies, taking local company sensitivities into account and tackling the cultural and trust barriers between the on-site office and the sector or corporate units.
- ○ Setting up the necessary infrastructure for large teams
  - ⇨ In large initiatives, it is very easy to forget some parts of the infrastructure. A checklist of things to keep in mind proves quite helpful and facilitates project administration.
- ○ Tracking progress and synchronization of all subinitiatives along common phases and milestones
  - ⇨ A customized integrated methodology provides the ability to compare progress across different initiatives.

An Integrated Team Member's Guide to Performing a Task

This chapter is written specifically for the team member. It explains how the team member can use the documentation that supports the product development process to understand and perform the required work. Program and functional directors should read this chapter to understand the details of the process.

6.1 INTEGRATED PRODUCT TEAM MEMBER RESPONSIBILITIES

The program and functional directors will give IPT members various assignments from the project plan, and this chapter will show a team member how the process documentation helps to complete them. Understanding how the project plan is organized is critical to allowing a team member to find and assimilate quickly what he or she needs to know and how to complete an assignment.

6.2 PERFORMING A SPECIFIC TASK

Performing a specific task means doing work that contributes to the development of the customer's product. The program director bases the assignment of these tasks on the program plan and his or her judgment of how resources should be allocated. Because the program plan is developed from the workplan templates, most assigned tasks are described in the process documentation. Completing a task involves

- ○ Understanding what the task involves, how it fits into the process integration framework, and to which customer deliverable it contributes
- ○ Understanding the individual work products that result from completing the task
- ○ Gathering the materials necessary to complete the task
- ○ Performing the task

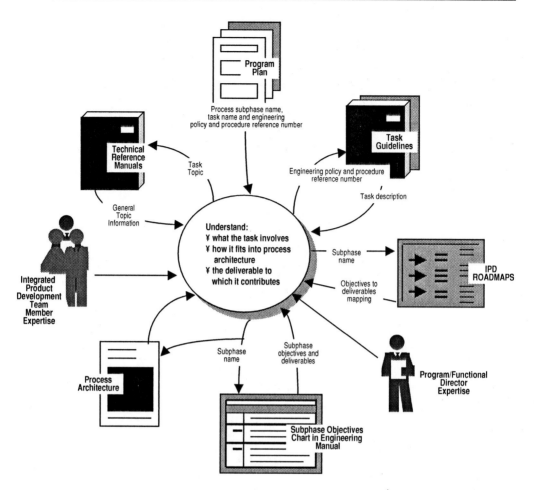

Figure 6-1: Process documentation used to gain an understanding of the task process framework and subphase objectives chart

6.2.1   Understanding what the task involves

The team members should make sure that they are well acquainted with the way tasks are organized (see the Chapter 3). Figure 6-1 shows the components that should be used to gain an understanding of the task, how it fits into the process framework, and the customer deliverable to which it contributes. These components are

- ○ The process framework and the subphase objectives chart from Chapter 1
- ○ Related objective and deliverables maps (ROADMAPs) that describe the workflow of deliverables
- ○ The engineering policies and procedures with guidelines and tips from the technical reference manuals
- ○ General topic information from the engineering manuals that are applicable only to certain tasks related to the topics covered in the manuals

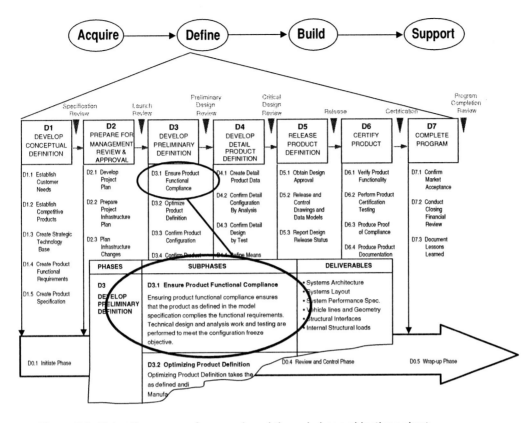

Figure 6-2: Using the process framework and the subphase objectives chart

(i.e., analysis, design, manufacturing, engineering, tooling, testing, and quality assurance)

As soon as the process framework has been documented by phase and function and the tasks have been described at the working level, the process documentation can be used to perform engineering work.

Look at the process framework and locate the subphase in which the task in question occurs. Read the objectives from the subphase objectives chart to understand the work involved in that subphase. On the same chart, see what customer deliverable(s) the subphase involves. A sample process framework chart and subphase objectives chart with customer deliverables is shown in Figure 6-2. From this information, a team member should understand how a task fits into the project and what the overall objectives of the subphase are.

Roadmaps

The related objectives and deliverables maps depict how, within a project plan, subphase objectives relate to the task groups, key deliverables, and customer deliverables (see Figure 6-3). This is also known as *workflow*. A separate ROADMAP for each subphase should be created.

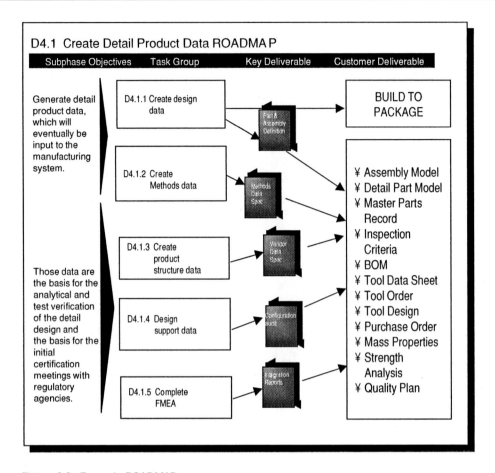

Figure 6-3: Example ROADMAP

Task guideline and design tip

To understand what the task involves, a team member should reference the task guideline and associated design tip (if one exists) from the engineering reference manuals. First, determine to which function the task belongs. Open the manual containing that function's tasks. Use the EPP reference number to find the guideline; numbers should be listed in ascending order. Figure 6-4 shows a sample design tip for a task guideline and also indicates the information that will help understand what the task involves:

- ○ Task name
- ○ Task description
- ○ Information for each workstep
   - ⇨ Workstep name
   - ⇨ Workstep description
   - ⇨ Work product name
- ○ Design tip

The task guideline will help team members find a design tips to help them better understand the tasks in question. Often, a design tip provides insight into the work involved, and it may occasionally indicate techniques that will be helpful in completing the task. A design tip may span several pages and can take many forms such as narrative, checklists, and diagrams.

The primary purpose of the task guideline is to specify what the task involves. It is not intended to explain exactly how to complete the task, though the design tip may provide some ideas. Each company should provide tailored engineering training programs, which, when combined with work experience, should enable

---

	ABC Company	
EP-200-1 **Engineering Practices and Procedures**	**DRAFTING REQUIREMENTS MANUAL**	Rev: A Page 1 of 12 Date: 1998-08-05
	Task Name: ITEM MASTER FORMAT	

TABLE OF CONTENTS

TASK DESCRIPTION: This Practice presents the basic Item Master (IM) sheet format and details the IM preparation. The basic IM sheet formats herein are for general CAD use for all drawings requiring a IM, either with or without a pictorial sheet.

Figure 6-4: Example task guideline and design tip

ABC Company		
EP-200-1	**ITEM MASTER FORMATS**	**Rev: A** **Page 7**

4.7.3   Annotation of PCD Type

Enter note numbers in column 19 in conjunction with part number entries arising from a PCD.

The corresponding note shall read "VENDOR ITEM – X" where X represents, as applicable, one of the phrases: "ENVELOPE DRAWING", "VENDOR ITEM DRAWING", "SOURCE CONTROL DRAWING" OR "SELECTED ITEM DRAWING".

4.7.4   Non ABC Company

For government and industry association standard parts and all proprietary parts, enter the complete part number and the associated code.

4.8   CANCELLATION OF DOMESTIC PARTS

Due to their permanent nature, do not delete the dash numbers of cancelled domestic parts from the PL table. Treat them as follows:

- Cross out part dash numbers that are no longer required for any "as delivered" configuration;

- Delete all other data in the PL table pertaining to the cancelled part. Delete from the PL all other entries (e.g. notes) uniquely pertaining to the cancelled part. See FIGURE 2.

4.9   SIZE

Enter the applicable basic size of the raw stock item from which each detail part dash number of the PL table is to be fabricated in column 13.

Examples of entries are shown in FIGURE 6. Items coded (1) are defined by industry standard stock sizes.

4.10   DRAWING NOTES

Enter drawing notes per Practice 7.0.1. Standard and Process notes are available in CAD, in LIBRARY, FAMILY, GENERAL NOTES.

4.10.1   Combine Notes

In the NOTES block enter the assigned note number from the PL table in a triangle.

The triangle coded note shall represent all the other triangle coded note numbers (in the NOTES) affecting the item, e.g.:

4.11   CONFIGURATION CONTROL TABLE

This element of the PL is optional. Delete the table if it is not required. Complete this table in accordance with the Configuration Management Handbook.

Figure 6-4: Continued

team members to perform tasks. The program and functional directors should recognize when the teams need help or training in particular techniques to complete a task.

As Figure 6-4 shows, the task guideline contains the name of each work product that results from completing each workstep. This example is a drafting requirements manual and explains the procedure for creating an item master file. If a design tip exists for the task, it may include additional information about the resultant work products.

EP-200-1	ITEM MASTER FORMATS	Rev: A
DATE: 1998-08-05		Page 12

MATERIAL TYPE	REQUIRED DATA (SIZE OR OTHER)	SAMPLES ENTRIES
BARS		
FLAT	THICKNESS X WIDTH	2.00 X 3.00
HEXAGONAL	WIDTH ACROSS FLATS	1.00 HEX
ROUND	DIAMETER	0.50 DIA
SQUARE	THICKNESS X WIDTH	2.00 X 2.00
BELTING	WIDTH X PLY	10(2 X 2 PLY)
CABLE		
ELECTRICAL	CABLE DESIGNATION	AN - 22
(MIL-W-22759)		
STEEL	DIA	1/8 DIA [1] 0.125
CASTING		
CONVENTIONAL	NONE	
BILLET	THICKNESS X WIDTH X LENGTH	2.00 X 3.00 X 7.00
CLOTH		
FABRIC	NONE	[1]
	TYPE	TYPE 1
	WEIGHT (OZ PER SQ YD)	6 OZ
WIRE	DIA OF WIRE X MESH	0.20 DIA X 40 MESH
CHORD	GAUGE	# 6 ga
COMPOSITE PLY	NONE	[1]
GLASS	THICKNESS	0.125
FELT	THICKNESS	1/8
FORGING		
CONVENTIONAL	NONE	
BILLET	THICKNESS X WIDTH X LENGTH	2.00 X 3.00 X 7.00
LEATHER	THICKNESS	3/64 [1]
	TYPE	TYPE II
PIPE		
STANDARDS	OD X WALL THICKNESS	1.00 OD X 0.040 THK

Figure 6-4: Continued

General topic information

In addition to the information that directly relates to a specific task, the engineering reference manuals present information about particular topics. For example, if a team member performs a drafting task, the drafting manual will help team members understand how the company performs that particular discipline and, hence, where the task fits into the overall process and what tools and techniques they may need to use.

The drafting manual is also a useful resource for some of the projects mentioned earlier such as preparing for a presentation or researching a particular topic for a task. Scanning the table of contents of each manual helps team members

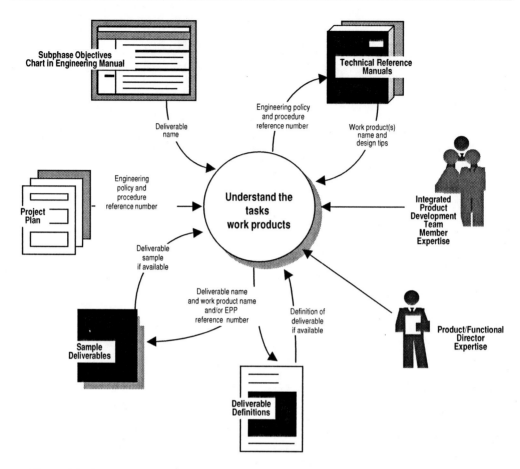

Figure 6-5: Components used to gain an understanding of individual work products

become familiar with what the manuals cover, and they will know when to reference the manual.

6.2.2   Understanding individual work products

Figure 6-5 shows the components that should be used to understand the individual work products that result from completing the task. These components are:

○ The task guideline and associated design tip from the technical reference manuals
○ The definition of the customer deliverable to which the task contributes
○ A sample of the customer deliverable to which the task contributes

Deliverable definitions

The deliverables architecture is described in Chapter 2. The deliverable definitions provide a table of contents and detailed definitions for each section of

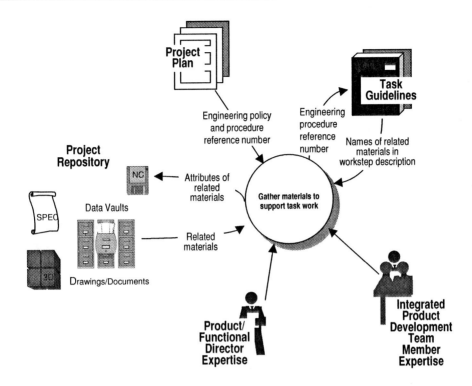

Figure 6-6: Components used to gather materials necessary to complete the task

each deliverable. If the work products from the task contribute to a customer deliverable, team members should refer to these detailed definitions to understand what is to be included. Sometimes work products will not directly contribute to a customer deliverable but are needed to complete a later task whose work products do appear in a customer deliverable.

To access information from the deliverable definitions, use the name of the deliverable to which the work contributes, the EPP reference number and the name of the work product. Each section of the deliverable is matched to its contributing EPP reference number(s) on the table of contents. Read the definition that relates to the work product(s), noting what is typically contained in it.

Sample deliverables

Sample deliverables contain samples from actual deliverables from previous programs. Deliverables vary with the needs of the program and customer, but if the work product contributes to a deliverable, then the team member should look at the sample copies.

The sample deliverables should be organized in the same manner as the deliverables definition (i.e., by deliverable and by section within the deliverable). Therefore, they would access them in the same manner – by deliverable name, engineering policy and procedure reference number, and work product name.

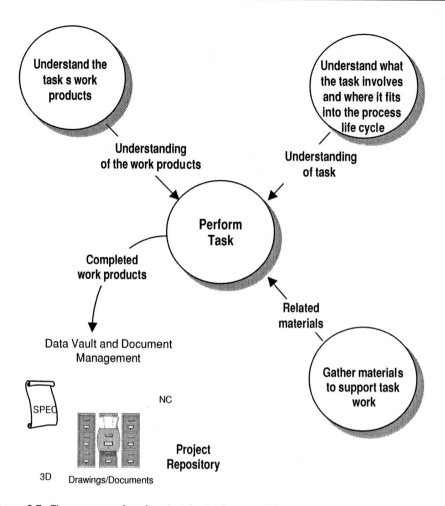

Figure 6-7: The process of performing the task on an IPT

From the information in the task guidelines, design tips, detailed definitions, and samples, team members should have a good understanding of the work products that result from performing any given task.

6.2.3   Gathering necessary materials to perform the work

Figure 6-6 shows the components that should be used to gather the materials necessary to complete the task. These components are

○ The related materials from the project repository, including the customer requirements, the design, the analysis, and the test reports
○ The EPP reference number from the project plan
○ The names of the related materials in workstep description from the task guidelines

- ○ The IPT member expertise
- ○ The program/functional director expertise

Related materials

The project repository contains all the project's completed work products. The repository can be stored in a Product Data Management (PDM) database or manually. This repository will be helpful if the team wishes to reference materials related to its work. For example, in creating a list of technical objectives, the team should refer to the requirements definition from an earlier task.

If the results of an earlier task are critical to the completion of the current task, they will usually be mentioned in the task guidelines or the workstep descriptions. The project repository may be assessed many ways – project plan number, sub-phase, deliverable, and product structure, among others.

6.2.4   Performing the task

Figure 6-7 shows the process of performing the task. A team member, in turn, should place the work products that result from performing a task in the program repository. Remember that the approach we describe in this book is merely a framework for planning engineering projects and a reference library for information about integrated product development tasks. It does not replace the experience and skills of the team member or IPT.

7 Program Structuring and Planning

In this chapter, we will examine an approach to top-down planning involving three principal activities:

○ **Confirming the program scope and environment**
This activity includes understanding the program objectives and scope, the technical and work practice requirements, and the cost and scheduling constraints. All of this is critical for product and functional directors.

○ **Structuring the program**
This activity involves defining the general approach to organizing and managing the program and developing a high-level program plan.

○ **Developing the detailed workplans**
This activity includes planning and estimating, at a detailed level, the outputs required to fulfill each of the program's subphase objectives.

We will also describe these planning activities and show the important linkages and considerations in using a top-down planning methodology. Figure 7-1 illustrates the results of these activities. The remainder of this chapter describes them in more detail.

Planning an advanced engineering program effectively is an essential prerequisite to delivering a quality product definition. With a firm understanding of the customer's business and a well-thought-out program plan, the product and functional directors can maintain perspective and a clear direction in the midst of rapid change. They can also remain flexible enough to accommodate requests for changes to the product's specifications or modifications necessitated by changing technology.

Program structuring and planning is accomplished by tailoring the Define process to fit the unique characteristics of the program. It defines the work that must be performed by identifying which tasks and worksteps in the workplan templates are required to complete the project. Clearly, the workplan templates, customer deliverables, and subphase objectives are interrelated. Any additions, deletions, or modifications to one will have an impact on the other two.

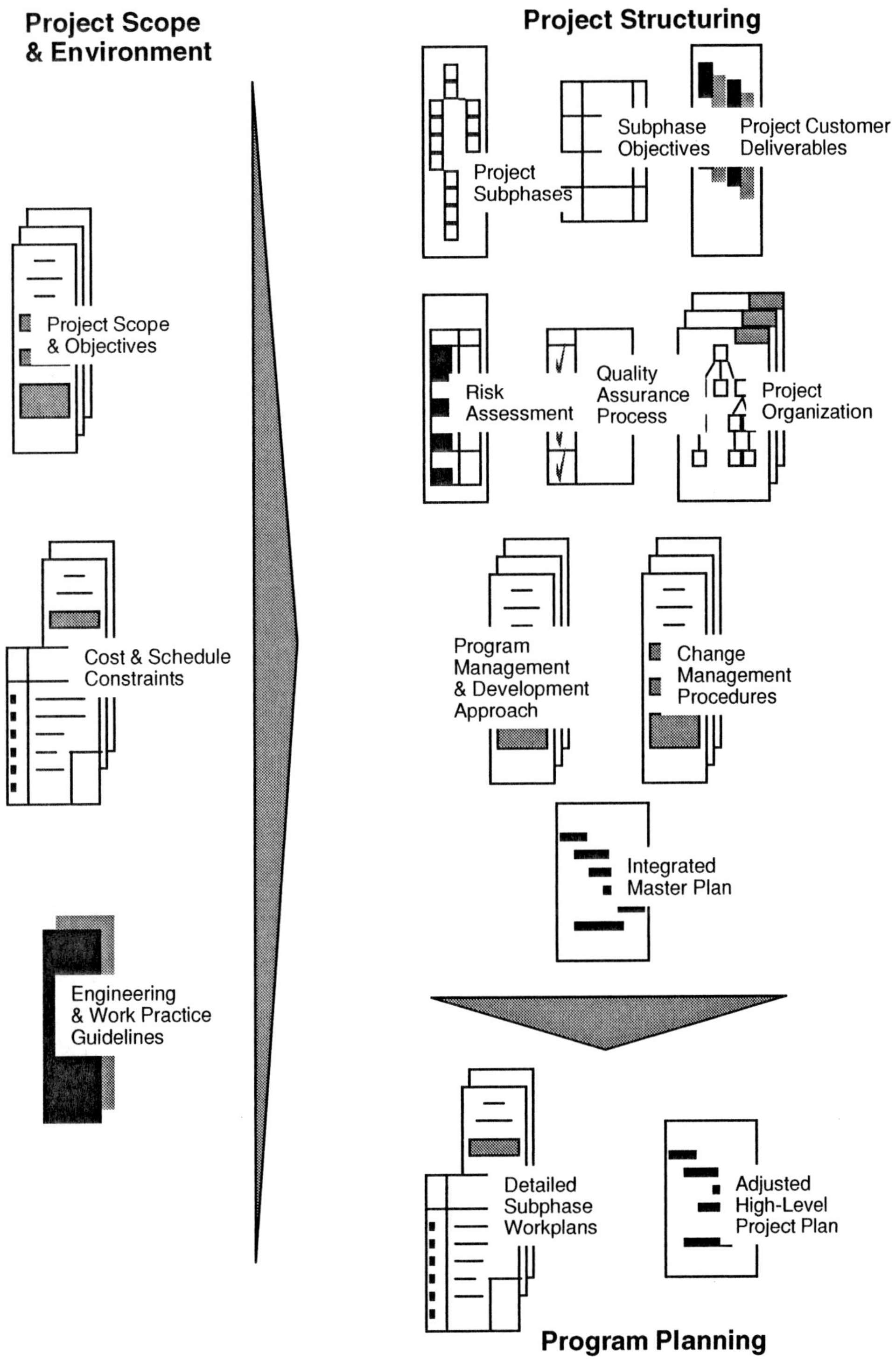

Project Scope & Environment

Project Structuring

Project Subphases

Subphase Objectives

Project Customer Deliverables

Project Scope & Objectives

Risk Assessment

Quality Assurance Process

Project Organization

Cost & Schedule Constraints

Program Management & Development Approach

Change Management Procedures

Integrated Master Plan

Engineering & Work Practice Guidelines

Detailed Subphase Workplans

Adjusted High-Level Project Plan

Program Planning

Figure 7-1: Approach to planning

A clear understanding of the program's scope and environment is a prerequisite for planning the program. The program's objectives and scope set the boundaries for the work effort and outline what is to be produced. The engineering and work practice guidelines define the procedures, methods, and tools that will be used for the program, including program management and system development techniques and reporting procedures. The scope should also include cost and schedule constraints and identify target milestones and resource expenditure caps.

This information is normally provided in the proposal, or Marketing Requirements and Objectives (MR&O). It should be supplemented with information obtained through interviews with corporate executives and the customer. In examining the scope and environment, you will need to focus on both known and unknown parameters [e.g., the program scope, the level of customer/partner support, and the Computer Aided Design and Engineering (CAD/CAE) technology environment] that will determine the program's resource and work practice requirements.

Confirmation of the program's scope and environment serves not only as a frame of reference for preparing estimates and workplans but also as a basis for assessing future changes in the program. The more clearly the environment of the program is understood, the easier it will be to structure and plan it. Confirming the scope and environment of the program with both the customer/partner and corporate executives is an up-front investment that will pay off over the life of the program.

7.1 PROGRAM STRUCTURING

Program structuring defines the approach taken to develop and manage the program. Through this effort, team members can gain a general understanding of the deliverables to be produced and the manner in which the program is to be run, before delving into the detail of creating an integrated master plan. Program structuring results in a high-level program plan (one that indicates the subphases and major milestones) and drives such critical detailed planning activities as modifying the workplan templates, estimating resource requirements, and producing a detailed schedule. Besides serving as a framework for creating detailed subphase workplans, program structuring outlines program procedures and standards (see Figure 7-1).

One of the main benefits of program structuring is that it enables you to better manage changes during the program. It does so by providing a clear indication of what the program's work will entail (e.g., its subphase objectives), which can be compared to and evaluated against any proposed scope changes (see Chapter 11).

Program structuring sets the stage for many of the key aspects of managing an engineering project. It is a key time to integrate the important elements of the program. The following topics are important elements of an engineering project. In fact, they comprise whole chapters of this book. However, program structuring and planning sets a cohesive orientation for all these elements on your projects. The elements are

- ○ Process subphases and deliverables (Chapter 2)
- ○ Risk assessment (Chapter 8)
- ○ Program reviews (Chapter 10)
- ○ Program management approach (Chapter 5)
- ○ Engineering change management (Chapter 11)

The orientation the program will take will be outlined in broad strokes upon completion of the program structuring process. For example, if your program is very complex, at the end of program structuring you will have defined the scope of the program and identified the risks that complexity causes. You will also have outlined a program review strategy to help minimize these risks, determined that you wish to manage the program very tightly because of the complexity, and decided upon the level of signing authority required to approve certain sizes of engineering change.

All these activities are interrelated, that is, the results of one could potentially affect the results of others. For this reason, program structuring is an iterative process. When performing these program structuring activities, they should be considered in the context of the entire program. One of the most critical linkages in program structuring is the one between engineering change management and program reviews. Utilizing the concurrent engineering approach is designed to eliminate unnecessary and untimely reiteration. The process integration framework is structured such that you reach agreement at a high level before expending resources on developing details.

It is best to define the subphases and deliverables first. The remaining program structuring activities (e.g., determining the program management approach, and tailoring the quality assurance process) need not be performed in any particular sequence, but all should be completed prior to developing any detailed workplans from the workplan templates. This is because the decisions made during program structuring affect not only the way existing tasks and worksteps should be tailored but also whether additional ones are needed. Figure 7-2 shows the major program structuring activities incorporated in the program management approach.

The outputs of the program structuring and planning checklists in Appendix D, the risk assessment questionnaire in Appendix E, the task planning process descriptions in Appendix C, and the quality evaluators selection worksheet in Figure 10-8 will help you make the transition from the structuring to the planning process. These worksheets allow you to note the decisions made during the

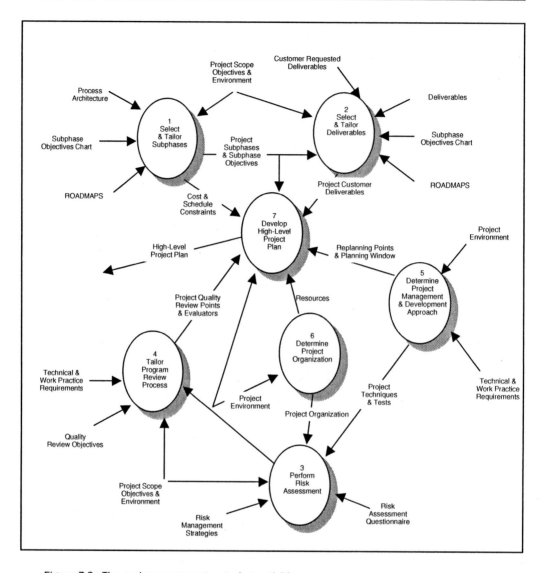

Figure 7-2: The major program-structuring activities

structuring process and to map those decisions to specific tasks in the workplan templates during program planning.

7.2 SELECTING AND TAILORING SUBPHASES AND DELIVERABLES

Determining a program's subphases and deliverables helps define the work to be completed and the results to be achieved during the program. Through this effort, the process integration framework is tailored at a high level in preparation for the detailed planning that occurs later. This tailoring consists of selecting and modifying and perhaps adding to the subphases, subphase objectives, deliverables, and sections of deliverables.

7.2.1 Tailoring subphases and subphase objectives

Determining the subphases that relate to your program and tailoring the subphase objectives results in a high-level identification of the work to be completed during the program. Through this effort, the work requirements are defined with precision, sufficient to let you link them to task groups in the workplan template.

Your choice of subphases and particular objectives is directly affected by the program's scope and objectives and by the existing program environment. These variables may sometimes require you to add a subphase or subphase objective. If, for instance, the customer wants you to take on the extra work of conceptual design of a wing (aerospace) or suspension (automotive), you would need to add an objective and related deliverables to accommodate this effort.

The process integration framework diagram, subphase objectives chart, and workflow diagrams will be helpful in determining the subphase and subphase objectives for the program. In addition, you may find it helpful to consult Chapter 2, which discusses the objectives, major task groups, key deliverables, and related considerations of each subphase.

7.2.2 Tailoring deliverables and sections of deliverables

Determining which deliverables result from the program work and which sections within those deliverables are appropriate to the program builds a framework for results-oriented planning. Through this effort, the deliverables are defined clearly enough for them to be directly linked to tasks in the workplan template.

If the deliverable definitions are used, each potential deliverable can be examined to determine whether, based on the subphase objectives and any requests from the customer/partner for specific deliverables, it should be produced. By examining the deliverable tables and detailed definitions for each deliverable, you can then tailor them to reflect the needs of your program (i.e., dropping unnecessary sections and adding new ones).

The workflow diagrams are also useful for tailoring customer deliverables, since they show how each subphase relates to key deliverables, customer deliverables, and task groups. Figure 7-3 shows a sample workflow diagram for the create detail product data subphase.

You can tailor the deliverables at any level of detail, from major sections to the most detailed subsections. Typically, decisions affecting major sections are directly linked to the chosen subphase objectives. Other factors, such as the technology environment or your specific program management and development approach (e.g., tools, techniques, and use of databases) will usually affect the more detailed subsections of the deliverable model.

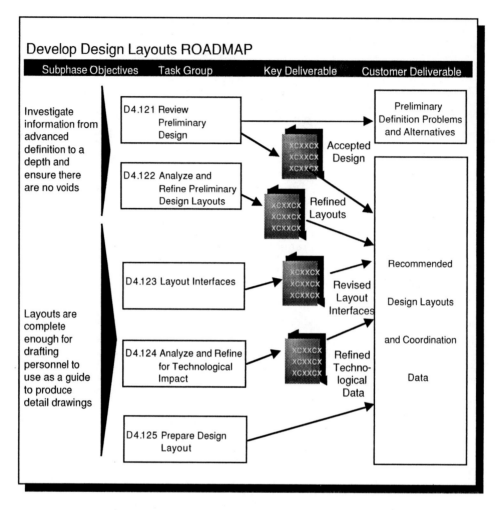

Figure 7-3: Sample workflow diagram for the create detail product data subphase

When reviewing anticipated deliverables, the customer may indicate that components of a deliverable are unnecessary. In such a case, take care to establish whether this is a mere packaging issue or a change in the definition (i.e., the scope) of the work. Before eliminating any work, you must determine whether it is work that is an essential prerequisite to proceeding with future activities and, eventually, to producing a quality product definition or work that falls outside the program's scope.

Once the deliverables are selected (see Figure 2-11), the sections and subsections should be noted in a table and linked to the workplan template (Figure 3.5). All subsections that reflect work requirements should be noted, with an indication of which sections will be included in the final deliverable given to the customer. These worksheets will be valuable later in the program, when the team has completed all other program structuring activities and is ready to begin program planning. Because they cross reference each deliverable subsection to the task in which it is finalized, these worksheets can be used as a starting point.

7.2.3 Performing risk assessment

The risk assessment procedure is described in detail in Chapter 8. As that chapter indicates, the principal tool for assessing risk is the risk assessment questionnaire. With this questionnaire as your guide, you can identify the program's risk factors and then note the responses on the summary sheet. You can then outline strategies for managing the risks that most threaten the success of the program. Recommended strategies for each risk factor are provided in the risk management strategy tables in Chapter 8. Figure 7-4 illustrates how the questionnaire ties to the strategy tables.

One deliverable of the program structuring process is to inform the customer, the partners, and the company executives of the program risks and the chosen strategies for dealing with them. Not only must these people be aware of the risks involved, but they may also need to take actions to help manage those risks. Of course, they may not agree completely with the team's assessment of the risks involved in the program, but it is essential that the various perspectives be reconciled. Only after a consensus has been reached about what risks the program faces can there be any agreement about what strategies to use to manage them.

7.3 PROGRAM PLANNING – THE INTEGRATED MASTER PLAN

Simply described, planning is an effort that helps to ensure the completion, on time and within budget, of all the work a program entails. An integrated master plan defines the work that will be done, how long it will take, who will do it, and when it will be completed. Adequate planning is crucial to controlling both the progress and quality of a program. Planning influences the program's progress by providing for continual reassessment throughout the life cycle. Although the initial IMP should be created before work even begins, the program should be reassessed and replanned at key points in the process framework.

Effective planning requires that you adapt the planning process to both the level of detail needed and the amount of information available. It is, therefore, often necessary to maintain planning data at various levels of detail. For instance, you may have enough information to create a detailed plan for the first two subphases but only enough data to plan the remaining subphases at a more general level. Two types of plans are provided in our approach.

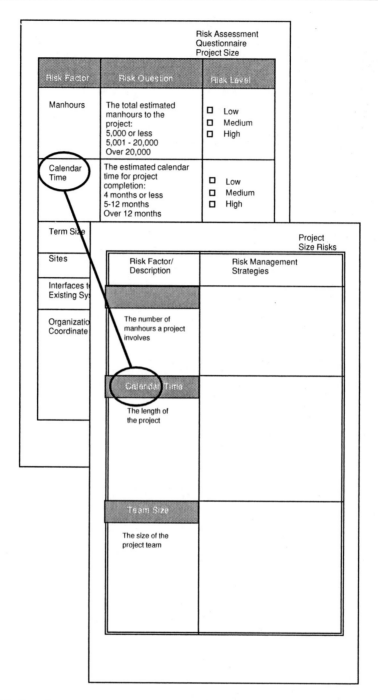

Figure 7-4: How the questionnaire ties to the strategy table

○ *Integrated master plan*

A plan that covers the entire program's life cycle and is summarized at the subphase level. As such, it identifies program dates and resource requirements for major milestones, such as subphases or customer deliverables. A

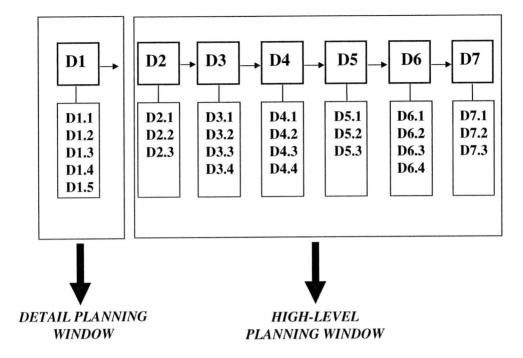

Figure 7-5: The planning window

natural outcome of the IMP is the integrated master schedule and program profitability analysis.

○ **Detailed subphase workplan**

A plan that covers activities whose outcomes are fairly clear. It is defined at a detailed level, typically at the task and workstep level. As such, it identifies each work assignment for a given period and matches the work to be done with specific staff members and calendar dates.

Planning at the IMP level provides a reasonable view of the overall program length and the approximate dates of key milestones. Although it can provide a quick estimate of how much time the remainder of the program will require, it should never be a substitute for short-term planning (detailed subphase planning), which must be done before work is performed.

The detailed subphase workplans require that only a portion of the work be estimated, assigned, and scheduled at a time. You will create a planning window; in essence, you will designate a portion of the program to plan at the detailed level. Planning windows can involve various pieces of a program, from a portion of a subphase, to a whole subphase, to multiple subphases. The planning window should represent the work about which enough is known to complete a detailed plan. This concept is illustrated in Figure 7-5. The size of a planning window should range from a minimum of three to four months to no more than six months.

Any more than six months' worth of data could produce inaccurate results because of change that can take place in, for instance, the availability of resources.

In addition to the two types of plans already described, some situations may occur where a mid-level plan that summarizes work at the task group level is required. A mid-level plan may cover only a portion of the process framework, typically a segment of work about which some detail is known, but not all the detail that is necessary to develop a detailed subphase workplan.

7.3.1 The workplans

The workplan is the program director's most valuable tool because it serves as the basis for delegating and managing work assignments. Like a conductor's musical score, the workplan provides the product director with the information necessary to orchestrate the efforts of the program team to meet short-term goals. It consists of individual work assignments that outline what will be determined and by whom. The workplan also arranges these assignments into an integrated master schedule, which accounts for task relationships and resource constraints. The personnel costs of work assignments and their associated materials are then calculated so that the program's total cost can be estimated.

Different workplans can be used to manage and control different aspects of a program and, therefore, are created during various stages of the program process framework. For example, a detailed subphase workplan may be created to manage the efforts of a specific subphase. The integrated master implementation workplan may be developed to manage the subphases (refine candidate configuration, confirm candidate configuration, develop manufacturing plan, develop tooling plan, etc.). A workplan should be created whenever resources must be closely monitored and managed to produce an end result.

The components of a workplan follow:

○ The assumptions under which the program will be conducted, including such issues as staffing, technology tools, planning, estimating, and scheduling
○ The deliverables associated with the program, such as the marketing requirement report and design requirements and objectives
○ Task and time estimates, which list all tasks associated with the program and provide an hour estimate for each
○ Estimates of the resource requirements needed to complete the work, including personnel, hardware, software, facilities, and other equipment and supplies
○ A schedule that presents a timetable for the program, including Gantt charts and reports that show the allocation of resources and the responsibilities of individuals
○ Estimates of the total cost of the resources needed for the program

Detailed workplans should be developed for any portion of the program whose work can be fully defined at the time the plan is created. A workplan provides the information necessary to assign the work and to monitor, evaluate, and control the performance of that work. This monitoring and evaluation may point to variances that call for adjustments in the workplan, work methods, or program approach.

Because workplans are the means by which program costs and schedules are reviewed and approved, it is important to make them as clear and complete as possible. Good workplans are both simple and graphic in nature. Their purpose is to identify the work, improve the program team's understanding of the work, and promote management review of the work; in brief, their purpose is to make the work visible.

As an organization gets ready to implement a new engineering process, the scope of the change will necessarily expand beyond the technical disciplines. The cost accounting system will likely also need to change. The numbering system that we developed in Chapter 3 provides a natural structure for collecting and analyzing costs. We have found that even old technology work order systems usually contain enough digits that will tie into the process framework and bill of material.

Teamwork is essential to the development of an attainable workplan and schedule. The collective experience and intelligence of all IPD team members can be invaluable when developing workplans and schedules.

7.3.2 Planning activities

Planning can be looked at as the sum total of the following eight major activities:

- ○ Tailor the workplan templates
- ○ Determine task relationships or workflow
- ○ Estimate tasks
- ○ Adjust estimates based on program risks
- ○ Assign resources
- ○ Develop an integrated master schedule
- ○ Tune the detailed workplan
- ○ Adjust the high-level program plan

These activities are shown in Figure 7-6, and each is discussed in the following section. A checklist for program planning is also included in this chapter.

Tailor the work plan templates

The workplan template (see Figure 7-7) can serve as a model for your program plan. Partner templates or special technology templates may augment these. Each template is a list of the tasks and worksteps typically performed on a program. The list includes estimating guidelines, estimating units, and skill type recommendations.

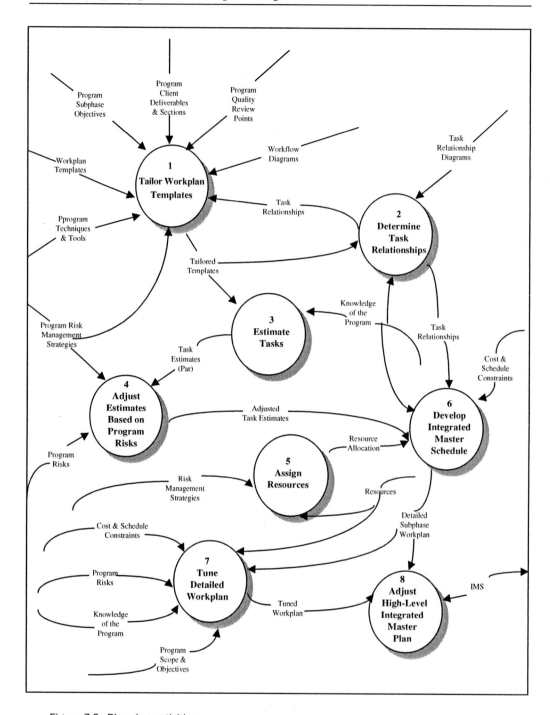

Figure 7-6: Planning activities

The templates provided must be tailored into an IMP that meets a program's specific needs. The decisions that were made during program structuring (e.g., subphase objectives, customer deliverables, planning and estimating assumptions, quality review points, risk management strategies) drive this detailed tailoring because they are directly related to the task groups, tasks, and worksteps offered in the templates.

To select task groups, use the workflow diagrams to trace the selected sub-phase objectives to specific task groups. In most cases, all tasks associated with a particular task group should be performed. If objectives were added, corresponding tasks groups, tasks and worksteps must also be added in accordance with the work that must be done to meet the objective. If objectives were changed,

	Risk Factor Categories						
	Project Size +	Project Structure +	Project Technology+	Additive Constant =	Total Weight x	Standard Estimate =	Adjusted Estimate
D0.1 Initiate Phase							
D0.2 Plan Phase							
D0.3 Execute Phase							
D0.4 Review and Control Phase							
D0.5 Wrap Up Phase							
D1.1 Establish Customer Needs							
D1.2 Establish Competitive Products							
D1.3 Create Strategic Technology Base							
D1.4 Create Product Functional Requirements							
D1.5 Create Product Functional Requirements							
D2.1 Develop Project Plan							
D2.2 Prepare Project Infrastructure Requirements							
D2.3 Plan Infrastructure Changes							
D3.1 Ensure Product Functional Requirements							
D3.2 Optimize Product Definition							
D3.3 Confirm Product Configuration							
D3.4 Confirm Product Financial Assumptions							

Figure 7-7: Worksheet for adjusting estimated risk factors

	Risk Factor Categories						
	Project Size +	Project Structure +	Project Technology +	Additive Constant =	Total Weight x	Standard Estimate =	Adjusted Estimate
D4.1 Create Detail Product Data							
D4.2 Confirm Detail Configuration by Analysis							
D4.3 Confirm Detail Design by Test							
D4.4 Define Means of Regulatory Compliance							
D5.1 Obtain Design Approval							
D5.2 Release and Control Drawings and Data Models							
D5.3 Report Design Release Status							
D6.1 Verify Product Functionality							
D6.2 Perform Product Certification Testing							
D6.3 Produce Proof of Compliance							
D6.4 Produce Product Documentation							
D7.1 Confirm Market Acceptance							
D7.2 Conduct Closing Financial Review							
D7.3 Document Lessons Learned							

Figure 7-7: Continued

corresponding changes in the task groups and deliverables must be made. We provide workflow diagrams as worksheets to assist in the tailoring efforts.

You can validate your selection of task groups by examining related tasks. Do this by referring to the definitions of the deliverables you developed during program structuring, in which each deliverable section references the task in which the work is finalized. Use this information to ensure that the tasks associated with the selected deliverable sections have indeed been chosen.

You may also wish to review all the tasks that cover a particular aspect of product development. In this way, you can add tasks at the appropriate place

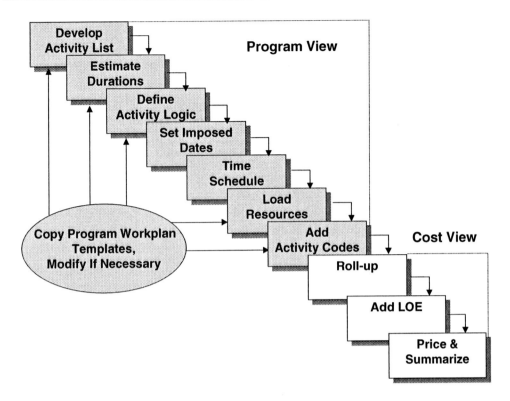

Figure 7-8: Developing a scheduled cost to complete

within the process framework. Aspect codes are used to categorize tasks within certain disciplines such as design, DFMA, and quality assurance (see Chapter 3).

The workplan templates are not a substitute for original thought in the preparation of a workplan. They should be customized by incorporating customer-specific nomenclature; resequencing certain tasks or worksteps; or adding, deleting, or expanding tasks and worksteps. See Figure 7-8, which illustrates where workplan templates are used.

Determine task relationships and workflow diagrams

After selecting and validating tasks through both the subphase objectives and the customer deliverables, the program or functional manager can assess the effect of adding, changing, or deleting tasks in the templates. These charts may indicate that a task not identified through the subphase objectives or deliverables may be essential to the successful completion of other tasks. Although these diagrams do not imply strict dependencies, they do aid in understanding how tasks relate to one another.

Estimate tasks

Like planning, estimating should be done at different levels of detail. The integrated master plan should, therefore, be accompanied by an order of magnitude estimate at the subphase level. Hours and costs may be allocated to a specific resource or a class of resources (e.g., stress engineer or mechanical designer) or by an organization (e.g., the company, subcontractor, or customer). Time can be represented as labor-days or labor-hours.

In detailed planning, however, the estimate should be very detailed at the task or workstep level. Conducting estimates at this detailed level will indicate the effort needed to complete each task or workstep. It is the translation of the work description (represented by the task or workstep name) into the manpower resources necessary to complete the work described. Use the workplan template to determine your workstep level "standard hours" estimates, taking into account the estimating guidelines and estimating units.

○ *Estimating guidelines*

Estimating guidelines are guidelines for the minimum, average, and maximum times typically needed to complete the workstep. Use these guidelines with care. They are not intended to substitute for experience as a product and functional director or for intimate knowledge of the program.

○ *Estimating units*

Estimating units represent the unit by which the workstep estimate should be multiplied to gain a realistic estimate for the entire work effort represented by the workstep. For example, the "develop standard time" workstep includes an estimating unit of "time per part" and estimating "guidelines per part." To estimate the requirements for all parts, multiply the estimate for the workstep by the number of parts to be analyzed.

The estimate should represent the effort taken for an average person of the skill type required to complete the work in an average environment. In other words, you should *not* be thinking about the specific individual who will be assigned the work when you initially estimate it. Often, there are shifts in personnel during a program that make estimating at the detailed level with particular people in mind too risky. Adjustments must be made at a higher level, such as the subphase or the entire program, through the program risk factors (see next section) to account for nonaverage resources or a nonaverage environment.

Adjust estimates based on program risks

What most affects the estimate is program risk. To accommodate these risks, you will need to adjust your estimates by a certain percentage. For instance, if there is a new CAD/CAE/CAM technology (virtual product development) to be used on the program, you may wish to increase the estimate for particular subphases by a certain percentage to account for the learning curve and inherent risks of implementing new technology. We have provided a worksheet for adjusting the estimated risk factors (see Figure 7-7). Use this worksheet to apply additional percentages to your base, or standards estimate. The input to this adjustment process should be the results of the risk assessment, which was documented on the risk assessment summary sheet.

Specific risks may affect only a few subphases. For example, the use of a new CAD system with feature-based modeling does not affect the prepare definition detail subphase, but it certainly must be taken into account during D4.1.1 Create Design Data and D4.1.2 Create Methods Data subphases. The worksheet allows you to apply risk adjustment percentages to specific subphases in the process framework.

Assign resources

Assigning resources means determining who will do each task or workstep. Specific individuals should be assigned to specific tasks or worksteps based on their skill type, skill level, availability, and concurrent activities. Often this is a point of conflict between the functional and program directors. In many organizations, the functional director has the final say.

Develop integrated master schedule

Scheduling translates resource requirements into elapsed time. This activity provides a method for balancing resources, coordinating tasks, and assessing the impact of change. When scheduling a program, tasks relationships, including strict dependencies, need to be taken into account. The workflow will prove useful during scheduling.

Developing an integrated master schedule entails four steps:

○ Identifying task dependencies
○ Identifying the critical path through the program
○ Identifying target dates
○ Backtracking through major milestones from the integrated master plan

Tune the detailed workplan

After a first cut at the detailed workplan has been completed, revisit the program scope, objectives, environment, and risks to fine-tune it. First, review the program planning assumptions that were documented during program structuring. Keeping the target date in mind, determine whether your plan will allow you both to meet the target date and to remain within the scope and objectives of the program.

The plan should also account for the specific concurrent product development techniques and virtual product development tools to be employed by the program team. Specifically, plans need to be made both to provide appropriate training in these techniques and to adjust specific estimates to accommodate their use. For instance, if you are using solid modeling techniques to design your parts, you should plan the effort involved in setting up the equipment and data files, training team members, and establishing the appropriate procedures for maintaining the software and data files.

Adjust the integrated master plan

After tuning the detailed plan, some adjustments to the IMP developed during program structuring may be necessary. Review the IMP against the detailed workplan to see if the estimate, schedule, or costs need to be adjusted.

At this point both the detailed workplans and the IMP are ready to be reviewed and approved by corporate and company executives. If the IMP has changed significantly, a review with the customer may also be advisable.

7.3.3 Building quality into the integrated master plan

Quality needs identified during program structuring will affect the selection, estimation, and scheduling of program tasks. Spending time and money on tasks that prevent errors from occurring is the surest, most cost-effective way of building quality into the program. Preventive measures, such as using a holistic approach, following program-specific procedures, and planning a program, are employed throughout the life of a program.

Program management tasks that lead into or follow segments of work, such as training, reviews, and testing, are normally estimated and scheduled as discrete tasks in the program workplan. The workplan templates identify the major quality review and testing tasks. These tasks can be expanded, added to, or eliminated, depending upon the quality assurance needs identified during program structuring.

A few tips on building quality into the IMP follow. More detailed information on the program review process can be found in Chapter 10.

7.3.4 Strategies and plans within the process integration framework

Planning for program reviews

The appropriate program review points can be selected based on the decisions made during program structuring. The duration of each program review depends upon the number of program review participants and the complexity of the deliverables.

When scheduling the program reviews, it may be desirable to schedule reviews of sections of the deliverable before the end of the subphase to minimize rework. This is especially effective when large chunks of work are being developed or when assumptions are not fully agreed upon. For instance, during advanced definition, you may wish to have a program review of the candidate configuration before beginning the design of manufacturing plans, layouts, tooling plans, and the like.

Planning for walkthroughs

Walkthroughs should be brief and focused. The work product being reviewed should be small enough to be examined in 30 to 90 minutes. If a product cannot be effectively covered in so short a time, it should be broken down into smaller pieces and handled in a series of walkthroughs.

Walkthroughs generally involve three to seven IPT members, one of whom is the producer, or author, of the work product. Be sure to assign participants based on their ability to contribute to the product's evaluation.

Because one work product can serve as either a reference or the basis for a subsequent work product, walkthroughs should be scheduled immediately after the completion of a specific work product. Timely reviews minimize the amount of rework required.

The worksteps for conducting walkthroughs are defined at many points during the process framework. These recommendations should be reviewed and walkthroughs added or deleted based on the role walkthroughs play in your program review process (see Chapter 10).

Planning for testing

When developing the initial program workplan, the program or functional manager must anticipate testing needs based on the program environment, size, complexity, and risks. Planning for testing starts early in the process framework, when a testing strategy is developed in the preliminary definition subphase. When planning and estimating your testing activities at the beginning of a program, you will have to rely heavily on your experience.

Be sure to build into the plan enough time to correct errors uncovered during testing and for the retesting necessary to ensure that those problems have been solved.

Planning for team training

The program environment, the product to be designed, and the management of the program dictate the IPT training requirements. Training should be scheduled to occur prior to the time team members will have to put their knowledge to use, but not so far in advance that they will have forgotten what they learned by the time they need to act on it. For instance, training in the use of CAD should be scheduled to occur no sooner than the refine concept subphase and not later than the beginning of the develop advanced definition phase.

Planning for program management

Allocating time for planning and managing the program is essential. Generally, planning, management, and administration activities absorb from 10 to 30 percent of the total program effort. A minimum amount of planning and management time is required for small programs or those that are well defined and threatened by only minimal risks. As the size, complexity, and extent of the program increase, the percentage of program effort expended on planning and management will rise.

The program management workplan template includes tasks that support all the work described in this chapter, as well as the work involved in program initiation, planning, execution, review and control, and wrap-up. Although some activities listed in this template occur at specific points in the process framework,

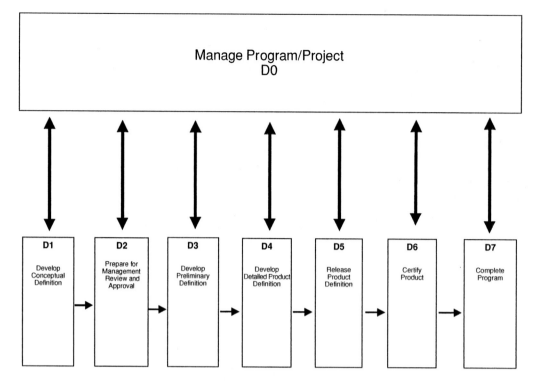

Figure 7-9: How the program management phase fits in with the process framework

many are ongoing activities and should be calculated as a percentage of the total program effort, rather than as a discrete number of labor-hours.

Figure 7-9 illustrates how the program management phase fits in with the process framework. Program management shown as DO cuts across all other phases and is applied to all phases of the life cycle.

8 Risk Assessment

In an engineering product development program, risk is the likelihood that the new product will not meet the customer's needs, be completed on time, or be completed within budget. This chapter focuses on these risks. It presents an overview of the risk assessment procedure and discusses the risk assessment questionnaire. Then it takes a detailed look at some risk factors common to engineering development programs and offers some strategies for managing them. The risk assessment questionnaire, summary sheet, and risk management strategy tables can be found in Appendix E.

8.1 PERFORMING A RISK ASSESSMENT

Advanced product development engineering programs are among the most complex activities conducted by mankind. They often involve millions of dollars of investment and thousands of hours of work by highly educated people. Risk is therefore high. Shareholders are willing to accept the risks involved in such an effort providing the potential returns are high. Yet even under the guidance of an experienced director, a program may fail because of inherent risks. By actively assessing a program's risks, however, you can reduce the likelihood of failure.

Risk assessment is a systematic method of identifying a program's risks and then determining appropriate strategies for managing them. It is also a mechanism for communicating the program risks to the customer/partner and to corporate executives. It sets the stage for working with the customer/partner to reach agreement on risks that are acceptable and in line with the organization's tolerance.

Risk assessment forces us to understand the problems presented by the program and to question the assumptions on which the program plan is based. It provides a set of techniques for evaluating the worth of alternative decisions and turns everyone involved into more informed decision makers. In short, risk assessment encourages excellent program management.

The results of risk assessment can help guide you in planning, estimating, and managing a program. For example, a specific risk may indicate that you need to plan for additional walkthroughs or program review points during the program or that you will need additional time to define product requirements. Perhaps a risk related to the use of new virtual product development technology may indicate that you need additional skilled staff or that additional team training should be planned.

The program or functional director is primarily responsible for assessing the program's risk. However, the customer and corporate executives also play an important role in helping identify risks. By independently completing a risk assessment questionnaire and jointly reconciling differing perceptions of risk, all parties come to the same understanding of the risks of the program and the appropriate management actions. During this reconciliation, everyone must listen and ask questions. Understanding clearly the potential risks will allow the project principals to identify them earlier should they materialize later on. Risk assessment is not the point in the program where the program or functional director takes on the worries of others. It is important to come to agreement on which risks the functional director will take accountability for.

8.1.1 When should a risk assessment be done?

The key to controlling program risks is identifying them early and then reassessing them throughout the program, using a systematic procedure. The earlier a risk is identified, the better the chance that it can be managed later.

At a minimum, a program's risk should be assessed during proposal development and initial program planning. It is also advisable to reassess the program's risk after a strategy has been adopted to deal with a high risk factor. This reassessment is to make sure that the action taken to lower risk in one area does not inadvertently increase risk in another. In fact, risk assessments should be performed whenever it is determined that the resulting information would be of value in managing the program. If properly identified and managed, risk should decline over the program's life. Figure 8-1 illustrates examples of risk assessment points.

8.1.2 The risk assessment process

The risk assessment process, as illustrated in Figure 8-2, consists of five steps:

- ○ Answering a set of questions about the program
- ○ Analyzing the completed questionnaire to determine areas of high risk
- ○ Determining strategies for managing the identified risk factors
- ○ Incorporating the strategies into the program management approach and into the program plan
- ○ Communicating the program's risks and the associated risk management strategies to the customer and corporate executives

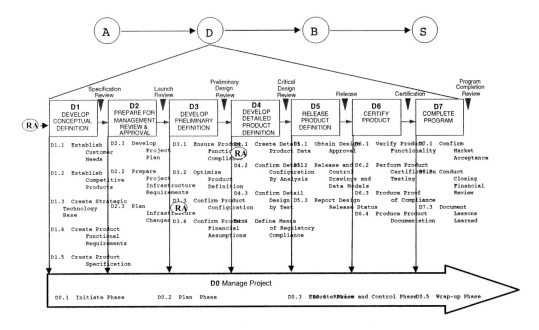

Figure 8-1: Sample of recommended risk assessment points

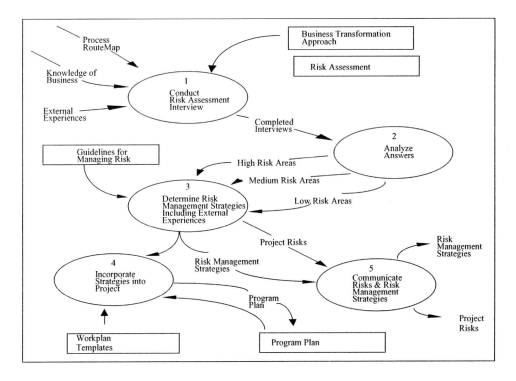

Figure 8-2: The risk assessment process

8.1.3 The risk assessment questionnaire

The questions in the risk assessment questionnaire (Table 8-1 in Appendix E) will help you identify and quantify overall and specific risks associated with a program. Each question offers multiple-choice answers, which, when scored, indicate a particular risk factor and its level of risk. The questions have been organized into three primary risk categories:

○ *Program size*

Program size risk is measured in man-hours of effort, calendar time, team size, number of organizations involved, number of locations, and the like.

○ *Program structure*

Program structure risk is measured in terms of clarity of program definition, sponsorship and commitment of the partners, customer, and so on; volatility of the program; type of program management approach used; and structure and qualifications of the team.

○ *Program technology*

Program technology risk is measured in terms of the new product's technological complexity and the program team's familiarity with a specific development approach and adaptation of new design technology.

Whenever possible, the questionnaire should be completed independently by the product director, partners, and corporate executives.

The purpose of the risk assessment questionnaire is to direct attention to specific program factors and to determine the risk level of each one. After reading the question that accompanies each factor, participants must assign it a high, medium, or low risk level. The results should be entered on the risk assessment summary sheet found in Table 8-2 in Appendix E.

The completed risk assessment summary sheet provides a one-page summary of the risk level for the program at a specific point in time. By focusing attention on any factor to which a high degree of risk was assigned, the program or functional director can take some action to lower that risk. Read the strategies for each of the program's high-risk factors to determine if one or more of them can be implemented.

8.2 STRATEGIES FOR MANAGING RISKS

Strategies to help manage specific risks are provided in Table 8-3 in Appendix E. Even though these strategies aid in risk management, they do not always enable a factor's risk level to be lowered. For example, if your program involves working with partners at ten sites, the risk for the "sites" factor is high. Using the strategies recommended for managing this risk to help reduce its impact, the number of sites will still be ten, but at least you will have altered your program plan and program management approach to deal with this fact. It is important, therefore, to remember is that risk is not necessarily bad, but the failure to identify and manage it certainly is.

Although most of the recommended approaches to risk management focus on specific aspects of a program (e.g., size, structure, or technology), some general strategies are helpful in managing a program's overall risk. These strategies are embodied in the process framework and are inherent in our overall program management approach. They include:

○ Having an understanding of the customer's expectations and clearly defining what will be done for the customer
○ Having appropriately skilled and trained staff assigned to the program, including management and technology specialists
○ Partitioning the work into manageable segments, a strategy inherent in the work breakdown structure
○ Examining alternative design approaches
○ Reducing the dependency of the program on other development efforts
○ Selecting appropriate tools and techniques
○ Involving the customers in the program (see "The Customer" in Chapter 4).

When developing a strategy to manage risk, examine the relationships and dependencies among risk factors. Sometimes a strategy to reduce risk in one area actually increases risk in another area. For example, a decision to reduce the risk of a lengthy program by increasing team size may also increase the risk of ineffective team coordination and communication.

8.2.1 Managing specific size risks

Large programs, by their very nature, are risky. Because of their size and duration, these programs present more opportunities for things to go wrong. For example, the program or functional director must continually be aware of and deal with risks that may never play a factor on a small program, the number of people on program team, the length of the program, communication among team members, multiple sites, multiple subprograms, and interfaces with other partners. You must carefully manage risks in these areas if the program is to succeed.

Several general strategies exist for managing risk on large programs. Formal program planning and tracking, for example, are essential on all such programs and should be supported by an automated program management tool. Regular program status reports should be submitted to both corporate and partner management. To accommodate their staff size, large programs also require mechanisms for communicating with all team members. Program intranets are ideal in these situations. In addition, changes in the business area and its requirements are more likely to occur over the course of a long program. In anticipation of this, change management procedures must be clearly defined at the beginning of the program. These procedures can also be used to expedite the resolution of issues and problems that arise during the program. Another large program strategy that may of value is the option of phased development, or phased implementation. Such approaches help reduce the risk on a large program.

In addition to these general risk management strategies, there are strategies particularly appropriate for specific size risk factors. The risk management strategy tables in Appendix E list these factors, along with the risks common to them and some of the strategies for managing those risks.

8.2.2 Managing program structure risks

Program structure risks represent a broad group of risks dealing with such issues as program definition, executive sponsorship and commitment, effect on the organization, program staffing, and program management. These risks often present themselves at the program's inception; therefore, they should be identified early. If the program or functional director fails to manage a program structure risk during the early stages of the program process framework, he or she may not have another opportunity to deal with it. For example, the lack of a partner or corporate program sponsor can easily be rectified during the early stages of the program. However, it may be virtually impossible to obtain a sponsor later on, especially if the program begins running into difficulties.

Several general actions should be taken at the beginning of any program to alleviate structure risks. First, encourage heavy customer involvement on the program. This includes identifying a strong customer program sponsor, recruiting an experienced program review board (see Chapter 10), and obtaining the active participation of the program team's customer members. Second, clearly define all deliverables. The customer must understand what work will be completed during the program. Finally, plan training early to help ensure that the program team has the necessary skills to accomplish the work.

In addition to these general risk management strategies, there are strategies particularly appropriate for specific structure risk factors. Table 8-3 in Appendix E lists these factors, along with the risks common to them and some of the strategies for managing those risks.

8.2.3 Managing process and product technology risks

Because it is always changing, technology poses a constant risk to advanced product engineering programs. Therefore, the program or functional manager must constantly be aware of changes in technology and understand how those changes will affect the program. These technological changes apply equally to changes in engineering methodologies and to the technology requirements of the final product.

New virtual product development (VPD) technologies can provide solutions to many problems faced by the program team. However, there are trade-offs to consider in choosing a new technology over the current, proven technology. A new technology may provide improved capabilities, but it may also place the program at risk if it is unproven, unfamiliar to the program team and the customer, not well integrated with other technologies, or hard to use. New technology is also continuously evolving in the products themselves. The use of the new product technology may be well worth these risks if failure to incorporate it could doom the product to technological obsolescence.

There are several general strategies for managing risks associated with the use of new or unfamiliar technology. For example, training needs must be identified and scheduled as early as possible. In addition, it is important for the IPT to include at least one specialist who has a broad technical background and has dealt with a variety of technologies. If such a person is not available for the team, the program or functional director should make provisions for outside technical assistance. The input of such a person is especially critical during the evaluation and selection of specialized product or process technology.

Another strategy for managing technology risk is to keep the IPT informed about all key design decisions through either regular team meetings or written communication. Developing a prototype is an excellent strategy that allows the IPT and the customer to gain some experience using the new technology before actually implementing it. This is often referred to as a solutions demonstration laboratory (SDL).

Program Initiation and Execution

This chapter deals with the important issue of "kicking off" a program. It is critical that a program begins in the right way. Having clear business goals and a professional work environment in place will provide a first-class atmosphere.

9.1 PROGRAM INITIATION

The start of a program is critical to its success because it is at this time that the project manager brings together the program team and set up the environment in which it will work. Because the IPTs are the most valuable program resource, ensuring that its members receive the proper training, a clear understanding of the program's goals, and the plans for achieving those goals are closely linked to the final success of the project. Recall from Chapter 8 that during project initiation you can resolve project structuring issues now better than at any other time during the program.

9.1.1 Establish program goals

Program and IPT goals should be established to ensure that the customer's needs are met. The first step in this process is to define the product functions and features, design-to-cost goals, and key milestones. Figure 9-1 illustrates the design-to-cost model. Notice how the integration of elements will drive the eventual cost. Investments in process capabilities can be as important to the final cost structure as the design itself. This model helps to establish the mission and focus of the IPTs. In support of the mission and focus, goals and planning milestones should be provisions for assessing progress toward them and a system for recognizing and rewarding those who achieve them. As program team members commit to these goals, it fosters a sense of team responsibility and accountability to their product and the customer.

The workplan template defines tasks to help you provide for the time needed to correctly initiate the program. These tasks are organized into the following three groups:

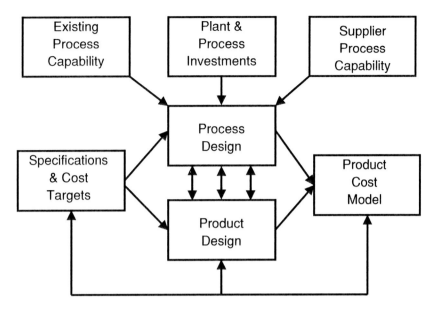

Figure 9-1: Design-to-cost model

○ Organize and brief IPTs
○ Conduct IPT training
○ Set up program environment

9.1.2   Establish product definition

The first step in initiating the product definition process is to understand fully the customer's needs. After they have been clarified, the product or system design requirements should then be allocated such that they satisfy these needs. The design-to-cost process is an integral part of this product requirement's allocation process. Analyzing systems requirements and evaluating alternate solutions will allow the team to develop a solution that meets cost and functionality requirements. This analysis is an integrated product team activity that allows all specialists (i.e., hardware, software, test, and manufacturing) to define how top-level requirements should be allocated to the system. Utilizing IPTs to conduct the trade studies prevents customer requirements from being misinterpreted. This process ensures that a solution is put forth that not only meets the customer's needs but also is technically and financially feasible and can be easily developed, tested, and built. The proposed design architecture is subsequently distilled into major products or subsystems in the form of a specification or requirements tree. Each product/subsystem is then evaluated to drive the product definition down to the lowest logical level.

These products or subsystems then become the basis on which to form the IPTs for the program. Specifications and tasks are then allocated to each product. These requirements allocations are also subsequently used to create the program work breakdown structure, statement of work, integrated master plan, integrated master schedule, basis of estimates, and team agreements. Maturity gates (see Section 2.5) identified in the IMP and timelines from the IMS can then be used to generate the earned value/cost management system utilized on the program. After the program IPT organization has been established, and a clear understanding of the work assignments for each has been identified in the form of a program plan (i.e., WBS, etc.), the IPT functional membership and resources can be assigned.

9.1.3 Organizing and briefing the program IPT

Organizing and briefing the IPT includes not only finalizing the staffing for the program but also briefing the IPT, the quality evaluators and program review board members selected during program structuring and planning (see Chapters 7 and 10). When briefing the IPT, the program or functional manager is most likely to bring them together as a group for the first time. He or she will capitalize on this opportunity to establish a common vision and understanding of the program and of the IPT members' individual contributions to it. This briefing is also used to show the commitment of corporate executives to the program and to fulfill a customer need, as well as to show customer commitment to the program. Therefore, plan to have the company executives, the customer, and the partner executives involved in both the program briefing and kick-off meeting.

Prepare the organization for integrated product development

One of the bodies of knowledge that sets the foundation for the holistic approach is integrated product development. Establishing an organization that is ready to implement IPD on a program requires the integration of project management and organizational behavior (see Chapter 13). Typically, the program structure is based on the work breakdown structure. In WBS, lower level products are assembled or integrated into subsystems, and these subsystems are then assembled or integrated into the customer deliverable. The IPTs are usually organized around the subsystems in the WBS. Their performance is the accountability of the IPT, but, ultimately, each IPT is responsible for the products' overall performance, from a customer and a business standpoint. The art of program management comes from allocating responsibility commensurate with the level of complexity and risk such that the team can manage. Typically, limiting the number of IPTs can be very effective for communication and coordination purposes; however, the number is clearly whatever best suits the individual end item and program needs. It should be stressed that regardless of the number of teams, a customer representative should be available to support the IPT and ensure customer needs are being addressed.

The IPT's outputs are coordinated and integrated across the program by an integration team. The IT is responsible for the overall technical, cost, and schedule performance of the program. The IT integrates, allocates, tracks, and manages the inputs and outputs of the program IPTs. The program manager typically chairs the program IT. As such, the program manager is ultimately responsible for the overall program performance from a customer, business, and technical standpoint.

When the program IPT briefing is completed, the team members should clearly understand:

○ The scope and objectives of the program and how it contributes to the customer's business strategy
○ The program organization
○ The high-level program plan, including subphase objectives and customer deliverables
○ The program risks and the chosen strategies for managing those risks
○ How the quality assurance process has been tailored for the program
○ Their individual contributions to the program
○ How they must work with one another on particular aspects of the program

The kick-off meeting plays an important role in program initiation. It will typically involve all team members assigned to the program. The program kick-off meeting signifies the official start of the program and is sometimes combined with the IPT briefing. The purpose of this meeting, however, is to focus on the detailed subphase workplans for the first segment of work on the program and to familiarize each team member with his or her initial assignments. The program or functional manager may also need to cover the following:

○ Program workspace arrangements
○ Organization of program files
○ Issue logging and resolution procedures
○ Engineering change management procedures
○ Other program-specific procedures and tools to be used
○ Progress reporting procedures
○ Scheduled IPT training

In Chapter 10, we discuss the particular needs of the quality evaluators and program review board members with respect to the program briefing.

Establish a resource profile

The integration of the program organization and WBS form the basis for determining the program resource requirements. The relationship is illustrated in Figure 9-2. For example, notice how subsystem B lines up with the B-IPT. Resources must be allocated to manage and fund the product development activity

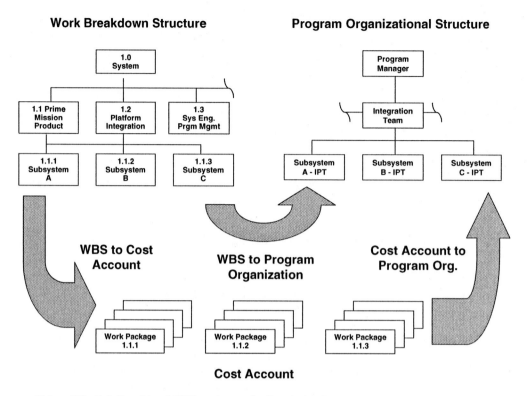

Figure 9-2: Relationship of WBS and organizational structure

effectively with the right people at the right time. Front loading of the program should be planned to support a do-it-right-the-first-time environment rather than paying to fix mistakes later in the life cycle. The overall program cost will be less if properly planned, executed, and tracked. The key is to base the allocation of people, dollars, and time according to the program design-to-cost goals, individual product requirements, and the assessed risk of the activity. Higher risk items should have a proportionally higher allocation of resources than lower risk items.

People In structuring work assignments for people, it is important to consider such aspects as having the right functional mix and skill level. Encouraging a barrier-free work environment that fosters the multifunctional interaction is essential to successful implementation of the integrated product teams. If the necessary skills are not available, then provisions should be made for appropriate training. Additionally, the program or functional manager must ensure that the personnel identified for the team are unlikely to be usurped by other programs or tasks.

Dollars The process of allocating dollars assigned to an activity or product is the next step. Tasks should be adequately funded and responsibly expensed. A tenet of IPTs is to drive authority and responsibility to the lowest level of the organization commensurate with risk (see Section 1.6). Responsibility to manage within a budget should be allocated to each team rather than by functional organization. Cost information should be collected and presented in a manner reflective of the individual teams. Allocation of the dollars within a team should be the responsibility of the team.

Time Time should be allocated to activities to ensure that schedules meet the needs of the product with the least possible risk. Up-front product life cycle planning, which includes all functions, suppliers, customer and end users, lays a solid foundation for the various phases of the product's life. When the events are clearly defined and understood, resources can be applied, and the impact of resource constraints can be better understood and managed. The more up-front planning that occurs, the less potential there is for unforeseen issues to surface later.

9.1.4 Conducting IPT training

To be effective, the IPTs need the skills necessary to complete their work within the original time estimates. Training in particular tools or techniques may be necessary to help team members build these skills. The program management workplan template includes steps for planning this training.

The first step is to review the product development approach and technical environment for the program. This information will help determine the IPT training requirements. Then the program or functional manager must lay out a training plan for the IPT and determine how the training can be accomplished. Technical training is available internally, through product suppliers (for software applications), on the Internet, or through local universities or consultants. To avoid the problems of overtraining or undertraining, a modular program can ensure that students receive the precise information they need.

Integrated product development and concurrent engineering represent a significant cultural change from the traditional product development environment. They require greater interaction among personnel and consensus decision making. Therefore, a key success factor is up-front education and training regarding these methodologies, roles and responsibilities in a concurrent engineering environment, and the use of IPTs. The entire program team (senior managers, functional managers, program managers, team leaders, and team members) needs to understand the basic approach and its benefits before the start of the program. Program offices are encouraged to use facilitators or trainers as much as possible and to enrich the training with as many local examples, case studies, and lessons learned available to improve understanding and to promote acceptance.

The program's customers and suppliers should be included as an integral part of the training. As IPTs are formalized, additional training that builds upon the initial training program should be offered. This supplemental training should provide detailed guidance on the implementation of the holistic management approach as it pertains to a specific team's product. It should focus on the roles, responsibilities, and interrelationships between the various disciplines within the team and between teams, on the participation of core and support team members, and on bringing the group together as a cohesive team. The training should identify the team's customers, both internal and external, and the customer deliverables. The training should be repeated for any new team members and as a refresher for other team members as needed.

Consensus decision making is a very important skill for successful IPTs. Many misconceptions about consensus will be overcome through adequate understanding and actual practice of the process. Training in consensus decision making should demonstrate consensus from multiple points of view to ensure that teams are able to apply it correctly. There is no conflict between consensus and functional excellence. The two can and should coexist. In fact, technically oriented functions learning to apply consensus for the first time often discover that creative designs can meet the needs perceived by all disciplines involved, thereby enhancing the ultimate end product design.

Training resources

Training is an "umbrella" term that includes formal classroom training, facilitated workshops, and other facilitated and self-directed activities.

Training framework Team effectiveness depends on both the technical and interpersonal skills of its members. Team dynamics training should provide the skills to work more effectively as a team. Teams may require training of specific individuals or an entire group. Training can vary from formal classes to helping teams identify and address specific issues. Program schedules and budgets should provide for training in the following key areas, as required to ensure adequate team performance:

- ○ Overview of the holistic approach
  - ➪ Definition
  - ➪ Roles and responsibilities
  - ➪ Process framework
  - ➪ Tools
- ○ Team leadership
  - ➪ Facilitation skills
  - ➪ Effective listening
  - ➪ Interpersonal communication skills
  - ➪ Effective meetings
- ○ Team building
  - ➪ Team basics
  - ➪ Team dynamics
  - ➪ Managing conflict

○ Problem solving
  ⇨ Establishing and meeting realistic goals
  ⇨ Managing time
○ Decision making

9.1.5 Setting up the program environment

A significant amount of time may be needed to set up the environment in which a large program will run. This effort includes

○ *Preparing the program workspace*
  This effort includes arranging for adequate facilities, office furniture, equipment, supplies and forms, telephone installations, and the like.
○ *Obtaining program equipment and supplies*
  This effort includes obtaining, installing, and testing the hardware and software needed to support the program's development and management tools as well as obtaining other program supplies.
○ *Establishing the documentation repository and program files*
  This effort includes determining how the files will be organized, setting up procedures for using them, and communicating the organization and procedures to the program team.
○ *Coordinating staff living arrangements*
  This effort includes arranging for temporary living quarters for program team members who are not local and arranging for transportation of these individuals from their living quarters to the program site for the duration of their involvement.

9.1.6 Staffing integrated product teams

Membership

Teams and teaming are at the core of the new process. They should be established at the start of a program. The teams should include key personnel from the proposal phase and should include a representative of each project stakeholder. Success rests heavily on the ability to build, empower, and nurture multidisciplinary teams. Considerable front-end planning and preparation should be utilized to assemble the right functional mix and skill level and to establish a barrier-free work environment that fosters multifunctional interaction. Team members should know how to complete the task. They should be focused on their specific product or process. Teams must clearly understand their roles and responsibilities as well as those of individual team members. They need to be empowered to make decisions commensurate with their skills and abilities and the level of program risks associated with their product. See Chapter 4 for a detailed listing of team members' responsibilities by program phase.

Integrated product team

IPTs are product-focused teams responsible for the technical, schedule, and cost goals for the development of a product and its processes. They are chartered to take into consideration the performance, producibility, maintainability, and eventual disposal of their product. Selection of the appropriate team members will depend on the unique characteristics of the product in terms of performance, cost, and risk. Any discipline that has a stake in the production, performance, and life-cycle support of the product should be represented. Multifunctional representation is necessary to ensure that timely, integrated decisions are made and that sequential processes are avoided. Team size will change depending upon what phase of the life cycle the product is in. IPTs will usually have two types of members: core and support.

Core members may be full-time or part-time, whereas support members participate on an as needed basis. Unique design, analysis, or fabrication activities constitute a support role. Support members from specialty functional organizations should be planned for on the IPT. Typically, support members participate in the IPT process as required by the IMP, IMS, and/or program milestones for their area of expertise. Selection of core team members can depend on product risk. A prudent course is to determine which functional areas present the greatest risk to the team's success and to select team members based on this information. A member is assigned to an IPT from a functional organization to represent the organization, provide the functional expertise, and perform tasks typically related to their discipline for the IPT. Core team members will typically include design engineering (hardware, software, systems), product support, manufacturing engineering, product/quality assurance, and the customer. Support members may include drafting/CAD/CAE, test engineering, contract administration, sourcing, purchasing, subcontracts, suppliers, reliability, maintainability, safety, environmental safety and health, human engineering, and finance. On collaborative projects, IPT members are also required for products or services developed by suppliers and subcontractors. This support ensures that both program and supplier/subcontractor concerns are visible and addressed when the team makes product decisions. These teams should be organized and should function under the same guidelines and methodology as other IPTs.

Integration team

Programs typically require systems engineering and system integration activities. In a concurrent environment, this is the responsibility of an integration team. The integration team (or integration teams, depending on the size and complexity of the system under development) is responsible for the overall technical and programmatic aspects of the end product or system being delivered. The ITs integrate requirements across IPTs, and they maintain the program and a product focus among teams. The IT serves both a system integration and a program integration function for the overall product or an appropriate subsystem of the product.

The IT concept is typically implemented with two teams and should be customer and program driven. When implemented with two teams, one team typically focuses on the overall technical performance of the final product or system (sometimes referred to as a system engineering integration team – SEIT). The other team focuses on the cost and schedule performance of the teams (sometimes referred to as a program integration team). In any event, the issue is not how many integration teams are used, but instead ensuring that system and requirements integration and overall program performance are coordinated among IPTs.

The top-tier IT is chaired by the program director and serves as a product integration function on the program. When an SEIT is used, it is typically chaired by the program technical director or chief engineer and serves as the system engineering and integration function for the program. In either case, the program director is ultimately responsible for the overall program performance from a customer, business, and performance standpoint.

The ITs should also facilitate the use of the process framework and manage IPT trade study efforts. They monitor the collective technical and cost direction of the program to ensure that performance, schedule, and design-to-cost goals are on track. The composition of the program integration team should include as a minimum the program manager, the program technical lead/chief engineer, the lower tier integrated product team leaders, the systems engineering lead, the customer representatives, the key suppliers, and the program manufacturing lead as applicable.

9.1.7 Team goals and agreements

Technical

The integrated master plan is the fundamental document for program coordination and control. It establishes the key program events and their associated tasks and requirements. In addition, the IMP structures the complete integrated effort necessary to develop, produce, integrate, verify, and deploy the product. In effect, the events, tasks, and requirements applicable to the program ITs and IPTs are coordinated, integrated, documented, and communicated as the basis for their team technical goals. As a team, the IPT members then use the IMP to establish their individual team and product technical goals in concert with those of the IMP. The IMP serves as the planning baseline and ensures the congruence of the individual IPT's goals.

Schedule

The integrated master schedule is the time-phased listing of the key events, their significant tasks, and the associated requirements contained in the IMP. The schedule for events and tasks applicable to specific ITs or IPTs are thereby communicated to them. They become the basis for determining their detailed product schedule. As a team, the IPT members should then establish their product

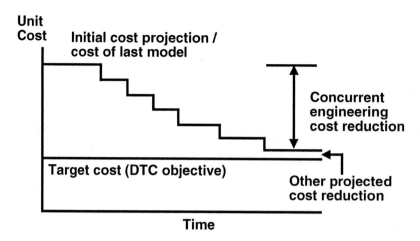

Figure 9-3: Achieving target costs

schedule in concert with the IMS. The IMS serves as the scheduling baseline and ensures congruence of the individual IPT's schedules.

Cost

Program cost goals are typically referred to as the design-to-cost (DTC) goals. Figure 9-3 illustrates how. Utilizing DTC, the team emphasizes cost targets early in the process framework to close cost gaps during proposal and concept design phases. They balance critical product requirements (weight, power, volume, and the like) against cost targets. The integrated product team concept is a key enabler for design-to-cost principles. All product team members supporting the program must participate in generating realistic cost goals and be committed to achieving them. Program DTC goals may address the complete life cycle of a product or may only address a specific phase or aspect of it (e.g., development cost, acquisition or unit cost, operation and maintenance cost, or disposal cost).

After a price to win has been established in the proposal phase, realistic program and product should–cost goals are established. The importance of cost realism in establishing these DTC goals cannot be overstated. Many programs have suffered from inadequate funding and were reduced or terminated because optimistic goals led to inadequate fiscal planning. These cost goals must then be elevated to the same level of concern as technical requirements and must be communicated to the ITs and IPTs as team goals. There must be a disciplined and a pervasive team sense of responsibility for cost control and containment. The product design should then converge on the DTC goal. The IT typically collects and tracks cost at a system or subsystem level, whereas an IPT should track and control cost down to a product's individual material and labor levels.

Team agreements

Contracts between the customer and the contractor ensure effective communication and clear understanding of the scope of work and the resources to be expended. Similarly, written commitments established between program directors and integrated product teams and approved by appropriate functional organizations provide the necessary framework for team performance. Written agreements with the IPTs help to reinforce and document the scope of work and resources defined in the customer's contract and to clarify agreements on roles, responsibilities, requirements, and constraints for each IPT.

These agreements should be established during the team formulation process and provide the basis for developing the detailed task descriptions and measurable performance standards for the team. It is important to remember to obtain consensus on the requirements, roles, responsibilities, scope of work, and accomplishment criteria imposed and implied on the IPTs within the program initiation phase. These agreements represent the law according to which the program director, IT, and functional managers will hold the IPTs accountable and the performance criteria according to which the IPTs will be judged and measured.

A team agreement also represents the written understanding among the members of the authoring team and cements the agreed scope of the team's responsibility and authority. It constitutes commitment to resources, schedule, and product definition between the parties involved. It is essential that the agreement be established with the concurrence and understanding of all affected parties and be kept up to date and readily available. As teams and requirements evolve or are refined during the program, the corresponding elements of the team agreements (scope, technical, schedule, cost) are revised to reflect current understandings.

9.1.8 Team operations

Colocation

Physically locating program team members together in a common, dedicated area with appropriate resources and facilities, is the best way to facilitate rapid and frequent information exchange among team members. Colocation (illustrated in Figure 9-4) can also help keep teams streamlined and focused and help minimize disruptive influences and counterproductive activities. It also serves to generate a sense of community, allowing teams to concentrate on finding solutions rather than describing problems. It does not require that all team members be present at the same location at all times throughout the program life cycle. Colocation implies placing all IPTs in neighboring dedicated areas, together with the IT and program director. Appropriate resources are needed to facilitate an effective colocation strategy. Virtual and part time colocation can be effective for support or specialty function team members, customer representatives, and suppliers. If individual teams cannot be collocated, technological solutions may be needed to facilitate communication.

In today's world where engineers work on $60,000 CAD workstations and peak program resource loads demand that contract and partner engineers work together, there is tremendous payoff to be gained from initiating a two- or three-shift engineering operation. A second shift allows for a lower capital base and therefore

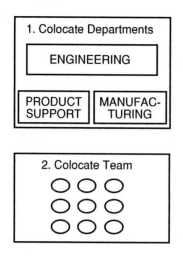

1. Colocate Departments	
ENGINEERING	
PRODUCT SUPPORT	MANUFAC-TURING

- More effective & timely communication.
- Greater opportunity for feedback & discussion.
- Better coordination.
- Less demanding infrastructure requirements.
- Rapid response to issues & process tasks.

Figure 9-4: Colocation of cross-functional teams

higher shareholder return. In a traditionally managed engineering organization, one entire discipline would be moved to the second shift under the watchful eye of a manager. In an environment with IPTs, the solution is to move one entire project or program to the second shift so that the IPT remains intact.

Communication

One of the most critical process elements for successful IPTs is communication. Open, barrier-free environments promote the exchange of information and ideas among team members and between teams. People, when given the environment, opportunity, and tools to openly exchange information and ideas, will often still hang back and look for visible signs that it is safe to do so. In successful teams, members have easy access to all pertinent program and product information. Team data are shared readily and frequently with other teams and those providing supporting functions. Many tools exist to facilitate the new approach by improving communication, teamwork, and productivity.

Meetings provide the forum for the most basic and direct form of communication and provide a good synergistic environment for stimulating proactive and creative exchanges among team members. Examples of commonly used meetings include design trades, design reviews, program reviews, and DTC status meetings.

Tools for team operations

Information needs to be received and processed in a timely manner to ensure the early identification of potential problems, and timely collaboration among team members to formulate integrated solutions. In addition to colocation, tools or enabling technologies make this possible (see Chapter 11 for a discussion of PDM tools). They enable cross-functional teams to share and integrate information and make informed decisions at the lowest level commensurate with risk. The Internet allows partners to collaborate to extents never before seen. Companies

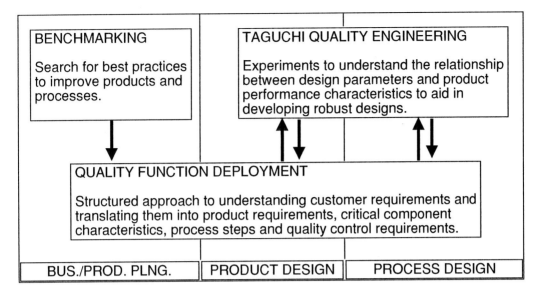

Figure 9-5: Quality design tools

A System to Ensure That Customer Needs Drive Product Design & Production

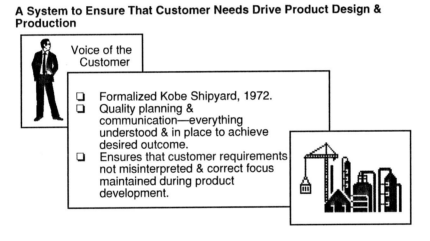

Figure 9-6: Quality function deployment

have learned that strong process linkages develop barriers to entry and increase the likelihood of future add-on contracts. The tools ultimately selected, however, should focus on simultaneous product and process development and program cost and schedule requirements compliance, and ensure that all necessary information is made accessible to all appropriate team members to perform against the program IMP and IMS.

Other process tools (illustrated in Figure 9-5) include various design, analysis, and management techniques such as quality function deployment (QFD – illustrated in Figure 9-6), design for test (DFT), Taguchi methods, design to cost, design for assembly (DFA), and variability reduction (VR).

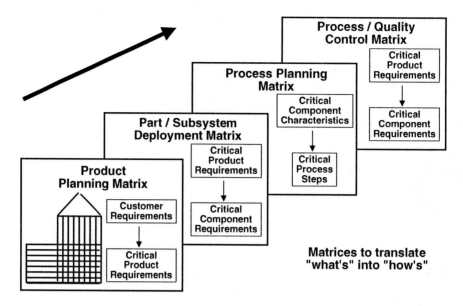

Figure 9-7: QFD planning structure

9.2 PROGRAM EXECUTION

The program director's greatest challenge is the ongoing monitoring and managing of the program so that it is completed on time and within budget and meets the program objectives. This activity includes following the plan, monitoring the program work, managing engineering change (see Chapter 11), managing suppliers, replanning the program as necessary and conducting, or coordinating meetings. We will examine these steps in detail.

9.2.1 Monitoring program work

It is the role of the program director and project engineer to know who is doing what and why they are doing it. IPT members can easily drift away from the program's goals, spending too much time on a specific task or misinterpreting an assignment. The program director must, therefore, constantly monitor the work to be aware of potential problems in assignments, team member interaction and communication, and new risks in the program. These situations must be dealt with in a proactive manner to help ensure that they do not throw the program off track. Program work can be monitored by applying the following guidelines:

○ Creating an environment that encourages the involvement and commitment of the IPT members in watching for and reporting potential problems
○ Being constantly aware of the critical path and how decisions affect it
○ Notifying the program executives immediately of serious problems that could have an impact on completion of the program or customer relations

○ Being aware of the relationships and interdependencies between the many parts of the program and knowing how changes in one program component will affect others

○ Establishing an effective progress reporting system that identifies issues, problems, and other special considerations related to the program

○ Adjusting assignments and the plan to correct problems that surface

9.2.2 Performance measurement

Performance measurement is applicable to all levels of activity and all phases of a program. It provides a cornerstone for assessing continuous improvement. To establish appropriate metrics for a program there are five steps:

○ Identify what you want to measure. These should be related to both the customer's and the program's needs.

○ Review existing measurement systems and generate new performance measurements if existing performance measurements are inadequate.

○ Make sure that the performance measurements selected are meaningful, valid, reliable, unambiguous, able to show trends, easy and economical to collect, and timely.

○ Select the proper tool for collecting, analyzing, and reporting the data.

○ Be sure to baseline your data so that you have a point of reference from which to determine the improvement. Finally, schedule periodic reviews of the data to assess where you are against where you want to be.

9.2.3 Progress reporting

The essence of all control systems is the ability to compare a plan or a budget to its actual results. Progress reporting is designed to provide this visibility at various levels and to various audiences, such as customer management, corporate executives, the program review board, the program's quality advisor, team members, and suppliers.

The reporting structure should provide up-to-date, useful program status information with a minimum of effort. If it does not, the reporting will prove cumbersome for team members and may not be done as promptly or as carefully as necessary. To guard against this problem, you should have a usable plan against which to track the program. This means not tracking at too detailed or too high a level.

During program structuring, some basic progress reporting decisions were made, as was a structure for this reporting. A particular program's levels, formats, and frequency of reporting depends on the customer's or the company standards, the automated tools used, the type of billing, the areas of risk, and the recommendations of the quality advisor. At a minimum, however, IPT leaders should report to you, in writing, at least once a week, possibly in combination with regular progress meetings. IPT leaders should convey the following information in these reports:

- Progress made (i.e., assignments – tasks or worksteps – completed)
- The amount of time spent on each assignment
- The amount of time left to complete each assignment
- The time spent on work not related to the program
- Assignments not completed on schedule and why
- Assignments planned for completion next period
- Problems, issues, or special circumstances that may cause delays or difficulties in the program

This information should be summarized for reporting to customer and corporate executives. Planned versus actual time reports and schedules can be developed and presented with this information.

The program director should actively monitor the program to

- Determine the degree of progress toward objectives
- Identify trends to take advantage of the constructive ones and correct harmful ones
- Note problems and issues so that they can be resolved quickly
- Note absences of coordination
- Note team members' failure to perform the tasks they have been assigned
- Anticipate problems so that corrective action can be taken early
- Be cautious of technical solutions that are pursued regardless of cost
- Manage the partners

In brief, program reporting information helps identify the issues, whose resolution will enable the program to progress smoothly and result in a successful implementation. To improve the efficiency with which issues are tracked and resolved, the program manager should establish an issues log at the start of the program and maintain it throughout the duration. This log is an excellent mechanism for communicating with the customer (especially the program review board), team members, corporate executives, and the vendor. It keeps issues visible and encourages their timely resolution. It also provides supporting information for the issues sections of deliverables. Information that should be captured on an issues log includes

- The date the issue is logged
- A description of the issue
- The date by which the issue must be resolved
- Its priority

○ Its status (if resolved, a description of the resolution should be included)
○ A list of actions necessary to resolve the issue
○ Additional comments

By accurately tracking the program and resolving issues early in the process, you remain in control and help ensure your program's success.

9.2.4 Managing engineering change

Three components of program management must be balanced throughout the program: scope, schedule, and resources. If any one of these elements changes on a program, one of the others will also change. The challenge of the engineering manager is to keep them in balance by being aware of any changes that occur in any one of the elements.

Because we were aware that most programs undergo many changes before completion, we addressed the need to deal methodically with such changes when we structured the program. During program execution, it is important to ensure that these engineering change management procedures are being followed so that perspective and direction can be maintained in the face of rapid change. This engineering change management mechanism also provides for dealing with conflicting demands from different user areas and to balance those demands with the program's schedule and cost constraints.

In Chapter 11, we will address engineering change management in concert with product data management. PDM provides a powerful tool to help managers cope with engineering change. It is a disciplined approach that enforces process discipline and ensures that all team members are working with the latest design configuration.

9.2.5 Replanning the program

In program structuring, planning windows and the replanning points for a program are determined. Replanning the program is very similar to the original program planning effort, except that we only review and update the program structuring information rather than create it from scratch.

During replanning, the detailed subphase workplans are developed for the next planning window. In addition, the integrated master plan should be updated to reflect any changes associated with the detailed replanning.

9.2.6 Conducting or coordinating meetings

Conducting meetings is an important aspect of managing the program. Various types of meetings require attention throughout the program, including quality evaluations, PRB meetings, meetings with partners, and internal progress meetings.

Meetings should be conveniently scheduled, with clear objectives and appropriate agendas that are followed. The minutes of the meetings should be issued promptly and should accurately document the results and action items. Completing action items on time is a responsibility of all team members.

Progress meetings with the IPTs provide an opportunity to motivate individuals, build team spirit, and monitor progress. Although we must deal with the details of individual assignments during these meetings, we should conclude them in a manner that refocuses the team on the big picture and the overall program goals. The integrated master schedule and the detailed schedule for the planning window should be reviewed at each meeting. Managers should be sure to distill the essence of complex problems, walk through unresolved issues on the issues list, and explain issues broader than the program that are occupying management's attention.

Customer and supplier involvement

Proactive, up-front teaming with both suppliers and the customer are critical to any program's success. It is even more important when using a concurrent design process. This process provides for rapid, low-cost, and low-risk solutions to be evolved jointly, with minimal risk of future reconciliations.

Customer participation and support is vital throughout the product definition process. It is the collective responsibility of the program director and IPT to meet the customer's requirements (cost, performance, and schedule). By involving the customer in this process as a member of the ITs, and as appropriate the IPTs, it ensures that the customer's needs are sufficiently and continuously being given proper consideration.

Key supplier participation in the program ITs and in program IPTs is also critical to achieving cycle time reductions and to optimizing data/communication exchange, as well as to ensuring that systems integration requirements are consistent. It also provides for additional cost visibility and cost containment at the supplier.

Specialty organizations support

As mentioned earlier in the IPT membership overview, whenever unique design, analysis, or technology activities are required in support of a program, members from specialty functional organizations should be called upon to support the appropriate IPT.

10 Program Reviews

One of the greatest benefits of applying process management to engineering operations is its ability to help build quality into the product development program. As mentioned earlier in this book, the methodology does this through a combination of people process and technology. It encourages user involvement and effective communication and applies techniques and results in complete documentation. In short, it helps focus on one of our most important goals: to provide high-quality engineering focused on customer needs. In this chapter, we look at the program review process and its role in ensuring product and process quality.

10.1 QUALITY ASSURANCE FACTORS

The quality assurance process helps ensure that work gets done correctly the first time, by assisting you in addressing those factors that are essential to a quality assurance (QA) effort. They include

- Focusing on the customer/partner expectations
- Doing the right task the right way the first time – First, concentrate on making sure that the right work is being done correctly, then focus on doing it faster.
- Following a proven approach to integrated product development, a way of building quality into the process and the product from the start
- Using proven program planning and tracking techniques – Framing a well made plan and then monitoring its execution is a way of infusing a program with quality from the start.
- Providing for training and education – Training is an essential part of improving quality in the product development process. The use of a consistent process framework will allow the company to transfer this knowledge to the employees consistently.
- Using structured techniques for analysis and design of the new product – The design for manufacture approach, supported by tools such as computational fluid dynamics (CFD), finite element modeling (FEM), and solids

modeling, provides a well-defined engineering approach to product development.

○ Developing a method to measure quality – Measurement cannot ensure quality, but it does offer a way to monitor the IPT's efforts to implement a high-quality system. Measurement can determine the product's ability to meet design requirements, the cost of achieving such quality, and needed improvements to the process.

○ Using a formal and rigorous testing process – Methodical testing leads to a well-engineered, high-quality product.

○ Using reviews to eliminate errors early – Make walkthroughs part of the program plan, allocating time for preparation by participants and for correction of any uncovered errors.

Although the entire define process is directed toward producing high-quality products, its program review process is the single most visible facet of the methodology's quality assurance effort. An explanation of this process, therefore, is the focal point of this chapter. First, however, to help you better understand the program review process, these pages discuss the concepts of quality and quality assurance in more general terms.

10.2 WHAT IS QUALITY ASSURANCE?

Quality assurance is a set of actions performed to bring about quality. This applies both to process quality and product quality. On an advanced engineering program, these actions are ultimately the responsibility of the program and functional directors, with the assistance of the quality assurance director. Part of the IPT's mandate is to work to ensure that quality is planned into a program, that it is engineered into the work products, and that it is verified through reviews and testing.

We identify seven elements of quality assurance:

○ Program quality assurance
○ Integrated product teams
○ Training
○ Program planning
○ Program-specific procedures
○ Reviews
○ Testing

Figures 10-1 and 10-2 illustrate the way these components interact to help ensure quality. Most organizations have some of them in place already. Our goal is to integrate them into a cohesive program.

To further comprehend quality assurance, it is helpful to understand how it differs from another key quality concept, quality control. Both efforts are

Program Quality Assurance

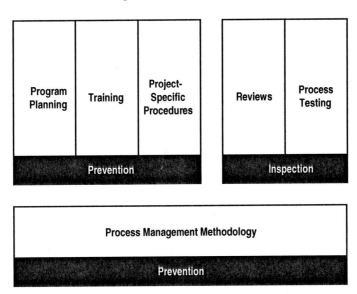

Figure 10-1: Quality assurance components

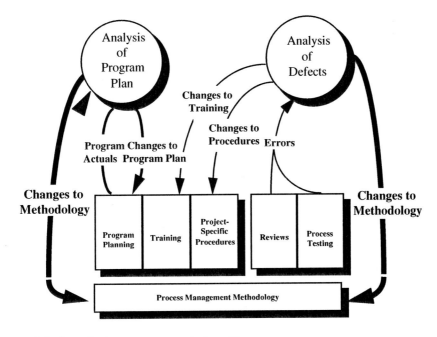

Figure 10-2: Quantitative measurement of quality

dedicated to improving quality, but one is reactive and the other is proactive. Quality control checks for and corrects defects in processes, goods, or services already performed. Quality assurance seeks to prevent defects from happening in the first place, through efforts built into the define/build process. Although quality control is not the same as quality assurance, it is one of the latter's most visible and frequently used processes. The principal reason for quality control, is that its data, or the feedback it provides, constantly steers the quality assurance program. It indicates how closely we are meeting the customer's needs and what areas we need to redouble our efforts to provide high-quality service.

10.3 DEFINING AND MEASURING QUALITY

Quality denotes a degree of excellence. We achieve it when we fully meet our customer's needs. Quality, therefore, is clearly a goal that guides us throughout the product development process. To meet the customer's needs fully, first we need to determine exactly what those needs are; then we must design a business process that meets them. Finally, we need to implement a process architecture that meets both the agreed-upon requirements and the customer's expectations.

Measurement is the key to improving quality because only through measurement can we determine whether an increase in quality has been realized. Although not all the components that make up high-quality design can be measured, two of its most important components can be measured: whether the product meets the agreed-upon requirements and whether the program is executed as planned (on time and within budget). Measuring quality through these two factors focuses on two sources of data.

○ *Process defects*
  Process defects are identified through review and testing sessions. Through careful discovery and recording of process defects, errors can be systematically eliminated to achieve deliverables of the highest quality.
○ *Program plan variances*
  Program plan variances are the difference between actual program execution data and that estimated for the plan. The program cost or schedule could exhibit variances. Analyzing these variances can better prepare the program or functional manager for problems later in the program and improve his or her planning and estimating skills for use on future programs.

These two types of data enable the program or functional manager and the IPTs to determine whether process quality is increasing during the course of a program or across programs over a period of time. More importantly, these measurements can be used to analyze the product development processes and methods to eliminate the cause of the defects and to reduce the number of future defects, not only on your program but on other programs across the company.

Three categories of measurement are illustrated in Figure 10-3. These time-based measurements impact corrective action and improvement.

A staffing ratio is a "ratio of full time equivalent design engineering team members to the full time equivalent cross-functional team member by function"

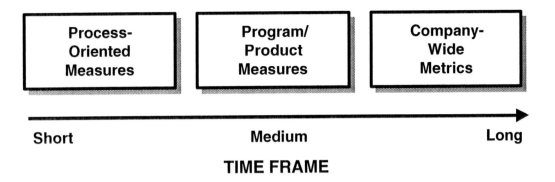

TIME FRAME

Figure 10-3: Measurement categories

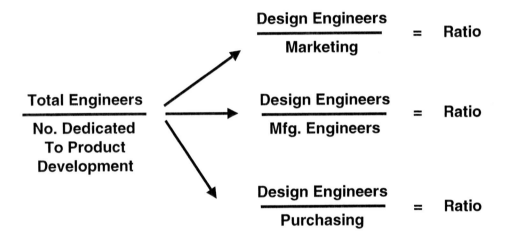

Figure 10-4: Staffing ratios

(see Figure 10-4). Product development is a cross-functional activity, and planning ahead will reduce resource bottlenecks.

Concurrency is "the degree of simultaneous team member activities during a given phase of a product development project measured in percent" (see Figure 10-5). The following example illustrates very little concurrency at the concept define phase.

We must link strategic objectives, performance measures, and improvement indicators with process improvement (illustrated in Figure 10-6). Performance measures and indicators are identified around processes.

One of our clients used a regular quality, cost, and schedule review very effectively. At these reviews, managers stand up and speak to the performance on these dimensions in their areas. They document problems with processes and then take responsibility to fix these problems. As a key element of a company-wide improvement project, it effectively saved the division from closure.

PHASE / FUNCTION	CONCEPT DEFINE	PRELIMINARY DEFINE	DETAIL DEFINE	RELEASE PRODUCT DEFINITION	CERTIFY PRODUCT
Mech. Systems	◨	■	■	◨	◨
Reliability	□	□	□	◨	□
Electrical System	□	□	□	◨	□
Structures	■	◨	□	□	□
Payloads	□	□	□	□	□
Stress	◨	◨	□	□	□
Mass	■	□	□	□	□
Aerodynamics	■	□	□	□	□
Flight Test	□	□	□	□	□
Tooling	◨	◨	■	□	□
Methods	□	□	□	■	□
IE	□	□	□	□	□
FAB	□	□	□	□	□
Assembly	□	□	□	□	□
Marketing	□	□	□	□	□
Customer Support	□	□	□	□	□
Q.A/QC	□	□	◨	■	■
R & D	■	◨	□	□	□

Figure 10-5: Concurrency activities

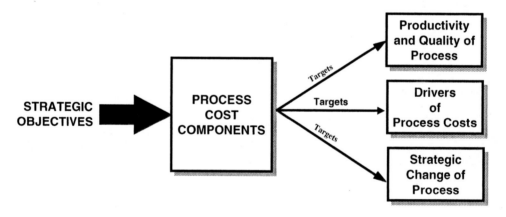

Figure 10-6: Linking of strategy to process improvement

The conceptual definition phase measures and indicators include

Performance measures
- Plan vs. actual define cost
- Cost per unit in production
- Research and development cost

Improvement Indicators (type)

- ○ Product design that wins the business. How does your company compare with competitors in bid cost?
- ○ Actual detailed build/support schedules and resource requirements
- ○ Changes from proposal plan/schedule to actual plan/schedule
- ○ Customer feedback and rating of quality deployment measurement
- ○ Maintenance of state-of-the-art information ahead of competition
- ○ Percentage of "product definition" areas supported by all systems
  - ⇨ Predict performance
  - ⇨ Indicate information
  - ⇨ Test for capabilities
- ○ Production development lead times
- ○ Research and development funding as a percentage of company revenue
- ○ Investment justification rating for R&D projects
  - ⇨ Tied to strategic plan
  - ⇨ Tied to operating plan

The gain management approval phase measures and indicators include

Performance measures

- ○ Plans vs. actual define cost
- ○ Cost per unit produced

Improvement indicators

- ○ Lead times for the Define process
- ○ Actual engineering hours vs. master schedule hours
- ○ Ratio of actual to requested team members
- ○ Number of master schedule changes vs. prior programs/projects
- ○ Completion of concept definition per original master schedule
- ○ Provision of right (qualified) resources on time and within budget
- ○ Personal/skill characteristics – teamwork, ability to change, and certifications
- ○ Percentage of contract with up-front funding

The preliminary definition phase measures and indicators include

Performance measures

- ○ Plan cost per unit
- ○ Cost of Define process/subprocess
- ○ Cost of quality

Improvement Indicators (type)

○ Production of quality design on time established in schedule (number Engineering Change Orders)
○ Lead time for major activities
○ Planned product support vs. actual by obtaining feedback from product support organization
○ Planned product/process definition vs. actual by obtaining feedback from build (producibility, etc.)
○ Obtaining customer acceptance on schedule and to quality specifications
○ Actual facilities/equipment specifications vs. planned
○ Percentage of contracts with up-front funding
○ Percentage of suppliers involved in design
○ Percentage of suppliers qualified
○ Percentage of parts analyzed for group technology; percentage of standard parts
○ Number of parts analyzed for make/buy
○ Average lead time by part, equipment, and material type

The detailed definition phase measures and indicators include:

Performance measures

○ Planned cost per unit
○ Cost of process/subprocess
○ Cost of engineering changes

Improvement Indicators

○ Lead time of Define processes
○ Bill of material accuracy
○ Routing accuracy
○ Number of engineering change orders within manufacturing lead time
○ Number of engineering change orders due to define errors
○ Number of parts requiring setup changes per assembly
○ Number of changes
○ Actual define time vs. estimated time
○ Design for manufacturability methods

The certify product phase measures and indicators include

Performance measures

○ Planned cost per unit
○ Cost of process/subprocess
○ Cost of engineering changes

Improvement Indicators (type)
○ Number of changes due to validation

The measurement of product development processes and activities is essential to achieving productivity gains and continuous improvement.

○ Metrics should initially be used to establish a baseline performance prior to implementing improvement.
○ It is essential to apply metrics at three levels: processes, product/project, and company-wide.
○ Product development must be viewed in a manner analogous to a factory. Projects are the unit of production, and company-wide design/build team resources define factory size.

Bottlenecks will result if the right ratios of resources do not exist between the various functional and department work centers.

10.4   QUALITY ASSURANCE THROUGH THE HOLISTIC APPROACH

The cornerstone of any quality assurance program in product development is the processes that guide the effort. When quality is engineered into processes, there is a base for all quality assurance effort. It provides a structured approach, appropriately skilled IPT members, performing quality reviews and measurement. It provides a set of stringent internal specifications to guide the way. These specifications include a high-level structure for identifying and planning major segments of work (a life cycle) as well as more detailed direction about what work must be done (task guidelines), who should do it (responsibilities), how to do it (structured techniques), and what will be produced (deliverables).

These components aid not only planning but also monitoring and controlling the work throughout a program. Perhaps, more importantly, this framework also includes activities that provide an opportunity to improve the quality of future programs.

10.5   TAILORING THE PROGRAM REVIEW PROCESS

You should plan for quality throughout the life of a program. Program structuring is the point to decide at a high level how to tailor the quality assurance process to meet the specific needs of your program (e.g., which review points are necessary and which types of evaluators are appropriate). The quality assurance requirements should then be defined in more detail and translated into specific review, testing, and training tasks. Individuals should also be assigned to specific quality assurance

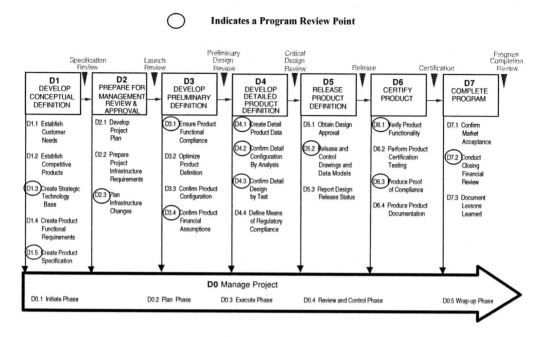

Figure 10-7: Sample program review points in the process framework

roles. During program execution, the program's program review processes are refined and replanned, based on the program's risks and the product's design. In collaborative projects, the quality assurance process takes on added importance. In a regular project, misunderstandings cause lost opportunities (and likely profit). In a collaborative project, misunderstandings cause lawsuits. Even though a PDM system (see Chapter 11) will help control the configuration of the product, rigorous program reviews are invaluable in helping control quality and the scope of the project.

During program structuring, this tailoring of the quality assurance process will involve

- Understanding what a program review is and how to plan for it
- Selecting the program review points for your program
- Selecting the types of evaluators required at each program review point
- Determining the involvement of the program executives
- Outlining the requirements of each program review component

10.5.1 Selecting program review points

In Figure 10-7, we suggest some program review points. The task then is to select which will be appropriate, given your particular program and customer. Factors that should influence your selection are the program scope, the objectives and environment, the risk management strategies that resulted from the risk assessment, and the selection of subphases and deliverables. Other variables that affect your decision should include the amount of customer involvement in the program, the complexity of the requirements, the clarity of the program's scope, the number of organizations involved, and the number of alternatives being considered. As these factors become more prevalent, increasing the frequency of the program reviews should be considered.

Based on this information, the program or project manager may choose to skip some of the recommended reviews, add others, or even change the timing of the ones recommended. For example, in the detailed definitions subphase, if the requirements seem to be conflicting or too vague, a program review of the requirements might be added before beginning work on the detailed engineering models. This minimizes the rework on the models, if requirements change at a later point. Though the workplan templates include a walkthrough at this point (D3.1), a more formal program review may be beneficial because of the complexity of the technical requirements.

The process framework can be used as a worksheet (see Figure 10-7) to document your decisions about the number and timing of program review points on your program.

10.5.2 Selecting the types of program review participants

After determining the appropriate program review points, the next task is to select the types of evaluators appropriate for each point. The individuals chosen to perform the reviews are known as quality evaluators. These evaluators are not IPT members but rather outside experts; they may be customer/partner employees or senior retired staff. Their task is to inspect project deliverables with regard to particular areas of expertise.

Quality evaluators generally fall into four major categories – design, customer, technology and manufacturing – but they may also include individuals from such specialized areas as materials, aerodynamics, propulsion, tooling, and manufacturing engineering in an aerospace company. The quality evaluators' general duties are

- To understand the deliverables they are responsible for reviewing
- To reserve adequate time to review the deliverables carefully
- To assess the correctness and completeness of each deliverable from its own particular perspective
- To participate in the evaluation process as though they were joint developers of the final product
- To write a summary of their findings from each quality evaluation
- To attend quality review board meetings when requested to clarify their comments

The quality evaluators should be chosen based on their expertise in at least one area important to the deliverable being evaluated. It is not necessary for each

	PERFORMANCE EVALUATOR	CUSTOMER EVALUATOR	TECHNOLOGY EVALUATOR	DESIGN EVALUATOR	AVIONICS EVALUATOR	MANUFACTURING EVALUATORS	MATERIALS EVALUATOR	MAINTENANCE EVALUATOR
D1.1 Establish Customer Needs								
D1.2 Establish Competitive Products								
D1.3 Create Strategic Technology Base								
D1.4 Create Product Functional Reqm'ts								
D1.5 Create Product Specification								
D2.1 Develop Project Plan								
D2.2 Prepare Project Infrastructure Reqm'ts								
D2.3 Plan Infrastructure Changes								
D3.1 Ensure Product Functional Compliance								
D3.2 Optimize Product Definition								
D3.3 Confirm Product Configuration								
D3.4 Confirm Product Financial Assumptions								
D4.1 Create Detail Product Data								
D4.2 Confirm Detail Configuration by Analysis								
D4.3 Confirm Detail Design by Test								
D4.4 Define Means of Regulatory Compliance								
D5.1 Obtain Design Approval								
D5.2 Release and Control Drawing and Data Models								
D5.3 Report Design Release Status								
D6.1 Verify Product Functionality								
D6.2 Perform Product Certification Testing								
D6.3 Produce Proof of Compliance								
D6.4 Produce Product Documentation								
D7.1 Confirm Market Acceptance								
D7.2 Conduct Closing Financial Review								
D7.3 Document Lessons Learned								

Figure 10-8: Recommended evaluators for the life cycle

evaluator to be able to review all aspects of the deliverable. The makeup of the evaluator team required at each review point, depends on the type of deliverable being reviewed or the special technology and integration considerations of the product being developed. In other words, not all deliverables require review by representatives from all quality evaluator categories.

During program structuring and planning, the program or functional manager must work with the customer/partner to determine the types of evaluators required and the specific individuals to fill these roles. They are based on the program's peculiarities and on the quality review guidelines. Then, once you have made your selections, improve your chances of successful reviews by making certain the evaluators know what is expected of them. (See Chapter 7 for help in selecting review points and evaluators, as well as worksheets to use in performing these tasks.)

Figure 10-8 illustrates the recommended evaluators for each review point. Figure 10-9 also illustrates the questions to be asked on a program.

The evaluators selection worksheet (see Figure 10-8) provides a means of documenting decisions about the types of evaluators for a program. Later, during program planning, the worksheet can be completed with the names of specific evaluators who have been nominated. This worksheet provides quick access to decisions when developing the detailed subphase workplans. On the form, the

Key Questions

> ➤ **Is the program being managed properly?**
>
> ➤ **Is the program developing/acquiring the "right" tools and processes to meet the program needs?**
>
> ➤ **Is there a master plan for the program and is it being followed?**
>
> ➤ **Is there a master schedule (with identifiable milestones) in place and is it compatible with program schedules?**
>
> ➤ **Is the program working to a documented set of requirements compatible with program needs and schedules?**
>
> ➤ **Do programs accept the requirements as valid and are the programs depending upon receiving the required tools and processes to meet their (the program's) schedules?**
>
> ➤ **Is there a realistic approach for deploying the tools and processes to the programs?**
>
> ➤ **Have the individual projects been properly prioritized?**
>
> ➤ **Is the program properly funded, including funding for deployment of processes and tools to the programs?**

Figure 10-9: Quality evaluators key questions

circles represent the recommended evaluators for each program review point. Simply circle those evaluators whom you plan to involve at each point in your program. Additional review points should also be noted on the form. There is space on the form to note the names of the evaluators when they are actually selected during program planning.

To fulfill their role, the evaluators need to understand the program, their role in evaluating deliverables and/or making business decisions, and the process framework. The quality evaluators and PRB members should be included in the regular program IPT briefing, but they should also be given a separate briefing that focuses on their particular responsibilities at each of the program's program review points. Their commitment to the program is just as important as that of the other team members. Items to cover with the quality evaluators and the PRB during this briefing are

○ The schedule of quality evaluations and PRB meetings
○ The extent of the commitment they are making as evaluators, or board

members and their general roles and responsibilities, including allocation of adequate time to review materials and/or prepare for meetings

○ Their individual responsibilities as specialist evaluators at designated review points (e.g., launch, PDR, CDR, release, and certification)

○ An overview of concurrent engineering, how it will be used on the program, and the concept of work products and key deliverables

○ The checklists available in the quality assurance monograph to guide them in their efforts

○ The sections of specific deliverables that each evaluator is expected to review (This may be explained separately to each evaluator outside the meeting.)

○ The evaluation notes or reports that you expect the evaluators to complete at each review point.

10.5.3 Determining the involvement of the program review executives

Our experience has shown that a review executive assigned to the program before work commences can be a valuable asset for the program. He or she should be experienced with the product technology and the major disciplines that the product entails.

The program advisor's involvement must be planned for, estimated, and built into the workplan. At times, it may be necessary to request additional program advisor reviews. This should be done if the program advisor has had limited exposure to the program's particular product technology. An increased frequency of reviews with the program advisor may be waranted during the preliminary definition and advanced definition phases so that the program advisor can determine whether the customer's needs and expectations have been identified and are clearly addressed by the plan.

10.5.4 Outlining program review requirements

To decide on what quality assurance actions to take, the program or functional manager must determine the role quality assurance will play in the program. This entails assessing the needs of the program and selecting the appropriate mix of quality components. The seven major quality components are

○ Program quality assurance
○ Integrated product teams
○ Training
○ Program planning
○ Program-specific procedures
○ Reviews
○ Testing

With the information obtained during program structuring, subphase objectives and deliverables, risk assessment, program management, and program organization processes, the basic requirements for each quality component should be outlined. For example, if the customer is able to commit only the minimum number of hours for quality reviews, you will need to make up for it in other areas. You could, for example, increase the number of walkthroughs, expand the role of the program review executives, or even advise the customer of the importance of program reviews, which may make them realize the value of providing evaluators. A checklist of program structuring and planning activities is included in Appendix D. This effort should also be planned and estimated on the program management workplan template.

10.6 ELEMENTS OF THE PROGRAM REVIEW PROCESS

Program reviews play a major role in the Define process. The methodology incorporates three different levels of program review:

- ○ Structured process and deliverable walkthroughs
- ○ Quality evaluations
- ○ Program review board meetings

Each level has a specified purpose and a defined group of participants. Their role is to examine the work produced by the IPT. Figure 10-10 depicts the relationships between these three levels of review.

10.6.1 Structured process and deliverable walkthroughs

A structured walkthrough is a short meeting at which several integrated product team members examine a work product with one common goal – to find errors. Walkthroughs occur at many different points in the life cycle.

Typically, a walkthrough is a formal, scheduled session that has a defined purpose and an expected result. As your IPT gains experience with these sessions, the process may become less formal. However, the goals remain the same, and the results should always be recorded. A walkthrough is the most powerful error detection tool during conceptual, preliminary, and detail definition. It is one of the few methods that can be used for inspecting the deliverables that are produced. A walkthrough can supplement testing in helping to ensure error-free results.

Although finding process errors is the principal goal of walkthroughs, they serve other general functions. For example, these meetings enable IPT members to understand each other's work. There is also an economic case for structured walkthroughs: the sooner an error is detected and removed, the less costly its elimination will be.

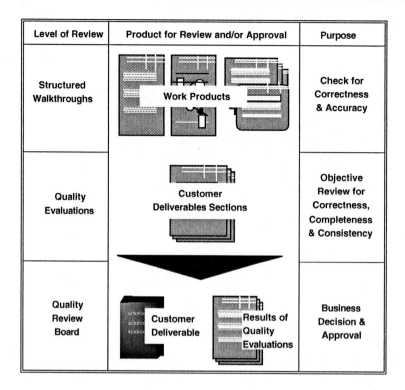

Level of Review	Product for Review and/or Approval	Purpose
Structured Walkthroughs	Work Products	Check for Correctness & Accuracy
Quality Evaluations	Customer Deliverables Sections	Objective Review for Correctness, Completeness & Consistency
Quality Review Board	Customer Deliverable · Results of Quality Evaluations	Business Decision & Approval

Figure 10-10: Three levels of quality review

The project budget should allow time for preparing and conducting the sessions as well as for revising the deliverables as necessary. Failure to do so can lead to a tendency to avoid other walkthroughs for fear that the errors they produce will cause additional program delays.

10.6.2 Quality evaluations

Quality evaluations represent the next level of program review. They take place at designated points at the program review points, and they assess customer deliverables.

We recommend program reviews at certain points throughout the product life cycle, and, as shown in Figure 10-11, the life cycle provides for these points. However, the actual number and timing of the program review points are joint decisions of the program and functional managers, based on the size, complexity, risk, and other characteristics of the program. In addition to recommending points at which quality evaluations are appropriate, the process documentation offers guidance as to what should be accomplished at these points. A comprehensive list of objectives for each program review point is provided in Figure 10-12.

In Figure 10-12, we illustrate the major program review points and describe their objectives.

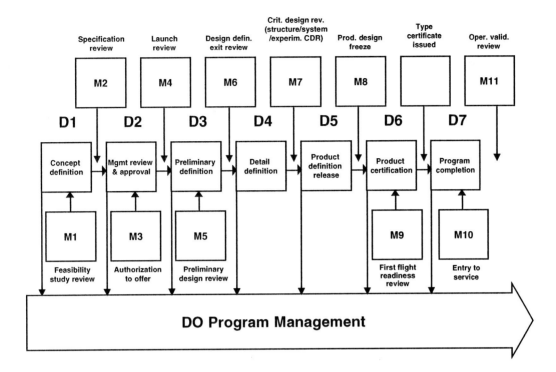

Figure 10-11: Quality review milestones in the life cycle – Example 1 – Aircraft

In Figure 10-13, we illustrate an aerospace engine manufacturer's toll gates. Toll gates represent a state of maturity that the deliverables must achieve before passing through the gate.

Like walkthroughs, quality evaluations are performed essentially to uncover errors. However, the effects of quality evaluations are much more far-reaching. Some of the purposes of these review efforts follow:

○ To help ensure consistency across a program's deliverables
○ To help ensure completeness of individual pieces of a program
○ To provide the final comprehensive review of a deliverable before it goes to the customer/partner for approval
○ To provide a much more independent review than that provided by walk-throughs because the experts who perform the quality evaluations are typically not members of the IPT
○ To help assess the deliverables from the customer's standpoint because many of these experts may be members of the customer/partner organizations

Although the process framework suggests certain program review points, the actual scheduling of evaluations is up to the program or functional managers. For example, they may find it advantageous for the quality evaluators to assess

Milestones	Definition	Context	Deliverables Completed at this milestone
M2 Specification Review	Specification review is the milestone at which the aircraft level specifications are reviewed to determine their capacity to meet the program goals.	In general, this milestone marks the completion of the conceptual design phase. At this point in the program the technical description of the product is developed in sufficient detail to proceed with preparation of the project plan for the overall project (i.e., schedule, cost, scope, etc.).	• Product Specification • Design Requirements & Objectives • Preliminary Master Lines • Conceptual Layouts
M3 Authorization to Offer	Authorization to offer is the milestone at which the conceptual design, business case and project schedule are sufficiently established so that the new product can be offered for marketing.	In general, this milestone marks the finalization of the technical description of the aircraft and the technical risk assessment has determined that a product offering is acceptable.	• Technical Description Documents • Top-Level Engineering Schedule have been selected
M4 Launch Review	Launch review is a milestone at which we review the readiness to launch the program.	By this time, the product configuration and its performance are satisfactorily known to a level that ensures that development will be achieved within schedule and cost and that the product will meet customer expectations.	• Business case defined • Suppliers have been selected In addition: • Resource Estimates • Statements of Work • Engineering Project Plan
M5 Preliminary Design Review	Preliminary design review is the milestone where the design is sufficiently complete to evaluate its acceptability to meet the technical requirements and cost, weight, and performance targets.	By this time the conceptual design has progressed to the stage of detailed layout drawings and the certification plan has been established.	• Master Lines Drawing • Detailed Layout Drawings • Interface Control Drawings • Certification Compliance Plan In addition • The Initial Board Meeting with the regulatory agency has been conducted
M6 Design Definition Exit Review	Design definition exit review milestone marks the end of the joint definition phase. A formal review is initiated to ensure that all preliminary designs are complete and that all suppliers are prepared to return to their facilities with their work share.	By this time the digital product definition is sufficiently detailed to facilitate the proper integration of all downstream efforts.	• Detail Definition Layouts and Interface Control Drawings are frozen • The aircraft 3D Electronic Mock-up is established

Figure 10-12: Comprehensive list of objectives for program review points – Example 1 – Aerospace company

Milestones	Definition	Context	Deliverables Completed at this milestone
M7 Critical Design Review	Critical design review is the milestone that marks the review of the final design. In general, all detail part, assembly and installation drawings are complete. The finalized design is reviewed against its acceptability to meet the technical requirements, marketing requirements, and cost, weight, and performance targets.	By this time the design has progressed to the stage of detailed drawings and advanced material orders have been issued.\n\nThe detail design is frozen following the CDR.	• Detail Part\n• Assembly\n• Installation Drawings\n• Loop 1 Loads\n• Stress Documents
M8 Production Design Freeze	Production design freeze milestone marks the formal release of engineering product data to manufacturing.	The basic aircraft design is now frozen and subsequent changes to any engineering product data now have to be handled via the formal change process. This is to ensure a controlled and managed release into the manufacturing system.	• Production Drawings\n• Nomenclature\n• Order Specifications
M9 First Flight Readiness Review	First flight readiness review milestone marks the readiness of the first flight test vehicle to fly.	In general, the aircraft configuration status, safety of flight issues, and testing status are reviewed.	• Experimental Permits and Authorizations are secured from the regulatory agencies
M10 Type Certificate Issued	Certification Issued milestone ensures that the design complies with the airworthiness standards and is acceptable for entry into service.	In general, all necessary flight & ground testing has been completed to guarantee the airworthiness of the aircraft, and all certification activity within engineering is complete. The certificate is granted by regulatory agency.	• Compliance Reports\n• Final Board Meeting\n\nIn addition:\n\n• The Approved Aircraft Flight Manual is complete.

Figure 10-12: Continued

deliverables as they are completed rather than wait until a formal review point. Such early detection of errors may prevent wasted effort in other areas.

10.6.3 The program review board

The program review board performs the third level of program review. The PRB provides one of the main links between the customer and the program team. Working closely with the sponsor, who is one of its members, the board is responsible

	CRITERIA TO PASS TOLLGATES
Tollgate 1	Identify the need for a new product, change or corrective action and scope options and anticipated resources. Evaluate business impact and need for a program.
Tollgate 2	Form team to develop concept for corrective action or new product program. Scope basic requirements, converge on options, and create preliminary plan including judgmental investigation.
Tollgate 3	Expand team to include manufacturing, sourcing, and partners. Generate an integrated business/technical plan. Obtain customer buy-in prior to proceeding. Define resources required to proceed with next phase.
Tollgate 4	Clarify technical and business aspects of the program, flow down customer requirements, refine costs, and obtain customer agreement.
Tollgate 5	Finalize technical and business aspects of the program, confirm manufacturing sources, program partners, and funding.
Tollgate 6	Appropriately expand team and create detailed program plan. Freeze design/process requirements, complete root cause determination, share details of corrective action/new design/process in determination for launch.
Tollgate 7	Launch program, finalize/colocate team and initiate next level of design. Refine tooling design, procurement, and planning activities. Proceed with analysis and tests to validate product/process change or corrective action.
Tollgate 8	Complete manufacturing planning, refine process definition, and procure hardware. Prepare customer documents as required.
Tollgate 9	Complete validation of corrective action/new product and conduct reviews as necessary. Submit required documents to external customers and agencies and obtain necessary approvals. Implement configuration management plans. Update and document internal practices/processes/procedures.
Tollgate 10	Produce product, deploy corrective action, and introduce to field. Collect manufacturing and field data to identify areas for improvement. Formally document any changes required for internal practices, processes, or templates. Set up periodic formal reviews to ensure that lessons learned are institutionalized and that new problems are identified, prioritized, and factored into the business activities.

Figure 10-13: Comprehensive list of objectives for program review points – Example 2 – Engine manufacturer

for approving the IPT's work throughout the program. More specifically, this body checks to ensure that the resulting product design meets their requirements and that it is developed according to predefined standards. It acts as a program steering committee, using the comments of the quality evaluators to make major decisions at each program review point.

The program and functional directors should work with the customer to select PRB members at the start of the program. They should be customer personnel who are executives in the business areas the program is serving. Each member must understand the technical areas he or she represents and the way it relates to other parts of the product. These members must also understand the objectives of the program and of their PRB. The PRB's overall objectives include

- ○ Helping to ensure that the product being developed and the process being used meet the customer requirements
- ○ Allowing all parties with a stake in the process and product a chance to express their concerns and suggestions
- ○ Initiating formal communication procedures between the various groups involved in the program
- ○ Helping to ensure that major organizational and operational decisions are made by a group of responsible executives, as opposed to one individual
- ○ Helping to ensure that all planned deliverables are created during the program and that they are reviewed and reworked as appropriate during each program review
- ○ Approving or disapproving the progression of the program through the life cycle, thus controlling the maturity gates

The board's principal function is to approve the IPT's work. The PRB's work, therefore, begins after the program or functional director and the various quality evaluators are satisfied with a deliverable and have presented the review reports at a PRB meeting. These sessions provide a forum for sharing information about the evaluations and for resolving any remaining issues. If the quality of the work is acceptable (based on the quality evaluators' comments and the resolution of issues), approval is granted, and PRB members can testify that the program is proceeding satisfactorily. If there are items of controversy, use the meeting as a means for discussing these items and for proposing an action program to lead to their ultimate resolution.

Although it includes reviews, quality assurance is more a preventative measure than a review process. By motivating and monitoring the business processes essential to providing high-quality design and developing high-quality products, quality assurance helps to ensure that these processes are working from the start of a program to its finish.

The quality advisor is a senior executive who oversees a program's quality assurance process. This person's role is a proactive one; he or she is assigned before the program begins, and his or her involvement continues through the postprogram review with the customer. As the title implies, the responsibility of the quality advisor is not to inspect but to counsel, to help ensure that the IPTs provide the customer with the best results possible. More importantly, this person will assist in building quality into the program, to help to ensure that the program is structured to meet the customer's needs and expectations from its outset. The quality advisor does this, providing the IPT with insight gained from involvement in similar programs.

More specifically, this person will work with the program or functional directors to

- ○ Incorporate quality assurance processes into the program plan
- ○ Make sure that the plan identifies and clearly addresses the customer's needs and expectations
- ○ Staff the program with personnel who have the necessary skills and technical expertise
- ○ Perform all necessary technical reviews
- ○ Develop a mechanism to identify and solve problems that arise during the course of the program

To maintain an objective view of the program, the quality advisor must be independent of the IPTs. He or she may not perform any other role in the program or report to the program or functional director. To qualify as a quality advisor, a person must be able to identify the critical factors upon which the success of a program depends. He or she must also have previous experience with the process technology and familiarity with the major disciplines the program entails.

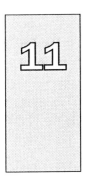

11 Engineering Change Management and Product Data Management

One of the most challenging aspects of an advanced product engineering program, especially from the program manager's perspective, is managing change. Although most changes are ultimately for the good of the program, some may seem frivolous, while others may cause delays that raise questions as to just how much they actually benefit the program.

A clear plan for managing change on any program will make managing it significantly easier. This chapter suggests an approach to engineering change management, in addition to providing insight into the concept of change as it pertains to integrated product development. One of the hottest topics in engineering management is product data management or PDM. PDM is a software application used to help manage the engineering process and related data. We will cover the key concepts of PDM and the relationship to process management in this chapter.

11.1 WHAT IS ENGINEERING CHANGE?

Engineering change is the addition to, deletion from, or modification of a product during its design, development, or manufacture. It is a normal occurrence during the progress of a program, most often caused by an evolving business environment or an advance in product technology. Regardless of the reason behind it, a change can have a serious impact on a program's scope, cost, or schedule.

Changes are not synonymous with problems. A *change* is an alteration to the specification or design of a product, whereas a *problem* is a variance between the expected result and the observed result. A problem, such as extensive vibration uncovered during testing of a product, is a defect that must be corrected. A change, on the other hand, can be evaluated before deciding on its incorporation into the product definition.

11.2 HOW TO DEAL WITH ENGINEERING CHANGE

On a large advanced product development engineering program (one year or longer), it is inevitable that a change in the market needs will affect product

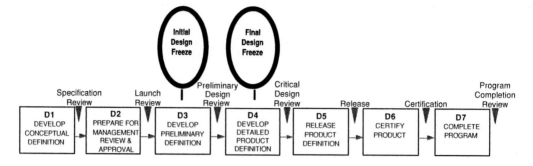

Figure 11-1: Design freezes during the IPD process framework

requirements. If the market is volatile, significant changes may occur in 6 months, and plans need to be made accordingly.

During the preliminary definition phase (Figure 11-1), changes to the requirements should be expected and incorporated into the specifications, as long as they fall within the program scope. Changes at this time may certainly require additional work, but it is far cheaper to include them at this stage, than to do so after the detailed definition phase is nearly complete. If a change request during the preliminary definition phase threatens to expand the scope of the program, however, it must be carefully evaluated, and its impact on the program must be determined. This is especially important if the program was bid at a fixed budget. You can use templates created during the project structuring to assess this impact.

Such changes may be inevitable, but they can be minimized. First, make sure that the marketing requirements and objectives analysis for the project is thorough. Many changes occur simply because of an inadequate effort on the part of the engineers to gather the product requirements. One strategy is to interview the customer engineers, those who have a solid understanding of where the market is going. This approach should not left to to sales and marketing personnel. The team's involvement in this step will allow the IPT to be able to anticipate changes instead of having to react to them.

After the preliminary definition efforts are complete, changes must be managed by adhering to prescribed engineering change management procedures. In this book, we document such procedures and provide insight into how they work.

11.3 MANAGING ENGINEERING CHANGE THROUGH DESIGN FREEZES

The process framework provides for two design freezes – the preliminary design review (PDR) design freeze and the critical design review (CDR) design freeze. A program should undergo no major changes after the PDR design freeze, which occurs after the develop preliminary support plans subphase. We suggest this guideline because major changes, for instance of the major component interfaces or materials change, would alter the fundamental product specifications of the project. Changes of this magnitude can greatly increase a program's budget and push back its schedule. Management should carefully review changes of this magnitude.

The final design freeze (CDR) occurs after the develop support plan subphase. Beyond this point in the program, no changes should be allowed unless they are reviewed and approved in the context of the formal engineering change management procedures. Figure 11-1 shows where these design freezes occur during the IPD process framework.

Prior to the design freezes, every attempt should be made to incorporate all reasonable changes that fall within the program scope. Every engineering change request (ECR), before or after design freeze, no matter how small, must be carefully evaluated because of the considerable time and expense it can cost the program.

11.4 THE ENGINEERING CHANGE MANAGEMENT PROCESS

The program management tasks help you to establish an effective change management procedure for your program. The intent is to enable product or functional managers to deal with change proactively, not reactively.

Timing is critical to an engineering change management procedure. If the detail definition phase (D4) is entered, without establishing a formal change management procedure, there is the chance that control of the program will slip away.

To establish the process, the product manager should meet with the customer/partner product director. Once the procedure is approved, the customer/partner, suppliers, and every member of the program IPT should be given copies. It is also a good example of data that should be contained in the program repository.

A viable change management procedure should include mechanisms for requesting, tracking, evaluating, approving, and implementing changes. The following discussion provides some guidelines to help product and functional managers address these issues.

11.4.1 Requesting an engineering change

Although most changes should be initiated by the customer/partner, any member of the program organization may submit a change request at any time by completing a change request form. See Figure 11-2 for samples of change request forms. Whether this form or another is used, it should capture certain information to help ensure that the changes are effectively evaluated and acted on. The form should contain the following information:

ENGINEERING CHANGE REQUEST FORM

Requester's Name:	Priority of Proposed Change: ❑ 1 Critical	TO BE COMPLETED BY PROGRAM DIRECTOR	
Program Name:	❑ 2 Important ❑ 3 Desirable	Date Received:	ECR No.:

Description of Proposed Change:

Reason for the Change:

References/Attachments: (Please list any documents or attachments which will aid in evaluating the change.)

TO BE COMPLETED BY PROGRAM TEAM

Reviewer:	Date:

Impact:

Estimated Hours:	Impact to Schedule:
Estimated Cost:	

Figure 11-2: Sample of change request form

○ Document/drawing/model

○ Requester's name and phone number.

○ Engineering change request number – This unique number is assigned to the engineering change request to help identify and track the change being requested.

○ Date received

○ Program/product name to which the change applies

○ Description of proposed change – This is a logical description of the change. It should be described in terms of what it seeks to achieve, rather than how it will be implemented.

○ Change class selection criteria of proposed change – This is an indication of the priority of the change being requested. Sample class categories from an aircraft example follow:

⇨ Before next flight, compulsory implementation. Essential for safety. *Reasons*: (a) Aircraft is grounded. (b) Design approval will be revoked by regulatory body. (c) Design approval cannot be obtained by regulatory body.

⇨ Time-limited compulsory implementation. Safety undermined. Implement before delivery. Service retrofit required. *Reasons*: (a) Operational limitations imposed. (b) Aircraft operates with deficiencies, no operational limitations imposed. (c) Noncompliance of an applicable airworthiness regulation; airworthiness directive may be issued by air transport authorities.

⇨ Production Implementation as early as practicable. Service retrofit optional. *Reasons*: (a) Design improvements – Does not meet specific customer, design, or type specification requirements; reduction in field maintenance, increased life, improved function or appearance. (b) Production Improvements – component impossible to purchase, fabricate, or assemble; production cost reduction; standards of workmanship unsatisfactory.

⇨ Record change, no physical effect on hardware. *Reasons*: (a) Typographical error, drafting omissions, clarification. (b) Engineering order incorrectly incorporated. (c) Reference information or callouts added.

○ Reason for the change – This explanation should anticipate and address challenges as to the legitimacy of the change. Whenever possible, it should include tangible benefits expected to result from the requested change, or real costs if the change is not made.

○ References and attachments – This includes a list of any additional documents that will aid in the evaluation of the change request.

○ Deliverables affected by the change – Deliverables that must be revised and reissued as a result of the engineering change request. A checklist of deliverables can be found on the back of the change request form (see Figure 11-3).

DELIVERABLES IMPACTED BY THE CHANGE REQUEST	
Check all that apply	
☐ Advanced Design Study Report	☐ MR&O Report
☐ Benchmark Report	☐ Strategic Technology Reports
☐ Five-Year Strategic Plan	☐ Product/Process Design Requirements and Objectives
☐ Design Standards Manual	☐ Drafting Design Manual
☐ Product Specification Document	☐ Product Requirements and Capability Plan
☐ Airworthiness Compliance Plan	☐ Technical Requirements Document
☐ Program Schedule	☐ Human Resource Plan
☐ Financial Plan	☐ Manufacturing Plan
☐ Program Partnership Plan	☐ MR&O vs Product Specification Matrix
☐ Business Philosophy	☐ Product Information Structure
☐ Work Breakdown Structure	☐ Preliminary Product Structure
☐ Database Structure	☐ Computing Requirements Plan
☐ Program Software Standards	☐ Facilities Plan
☐ Program Specific Business Processes	☐ Infrastructure Plan
☐ Configuration Management Plan	☐ Systems Architecture
☐ Systems Layout	☐ Systems Performance Spec
☐ Structural Lines and Geometry	☐ Structural Interfaces
☐ Internal Structural Loads	☐ Matrix of Preliminary Definition vs Product Spec
☐ Test Reports	☐ DFMA Worksheets
☐ Product Measures	☐ Trade Study Documentation
☐ Preliminary Design Report	☐ Updated Preliminary Product Definition
☐ DFMA Reports	☐ Updated Program Plan Report

Figure 11-3: A checklist of deliverables – Aerospace example

DELIVERABLES IMPACTED BY THE CHANGE REQUEST	
Check all that apply	
☐ Test Reports	☐ Part and Assembly Definition
☐ Methods Data Spec	☐ Vendor Data Spec
☐ Configuration Audit	☐ Integration Reports
☐ Matrix of Detail Definition vs Product Specification	☐ Test Reports
☐ Draft Compliance Plan	☐ Analysis of Preliminary Definition
☐ Design Authority Review Package	☐ Signature of Design Authority
☐ Preliminary Release Package	☐ Nomenclature
☐ Process Sheets	☐ Assembly Manuals
☐ Tool Design Drawings	☐ Drawing at Revision Status
☐ Physical Configuration Audit Report	☐ Component Test Reports
☐ Aircraft Test Reports	☐ Systems Test Report
☐ Certification Test Report	☐ Vendor Qualification Reports
☐ Approved Mod Sheets	☐ Certification Report
☐ Statement of Compliance	☐ Operating Manual
☐ Quick Reference Manual	☐ Service Bulletins
☐ Flight Manual	☐ Maintenance and Repair Manuals
☐ MR&O Compliance Matrix	☐ Change Request Documents
☐ Lessons Learned Report	☐ Budget Structure
☐ Parametric Estimates and Forecasts	☐ Technical Requirements Document
☐ Engineering Assignment	☐ Statement of Requirement
☐ Actual Hours vs W/O #	☐ Business Philosophy
☐ Quality/Cost/Schedule Visibility Charts	☐ Engineering Change Request Work Statement
☐ Closed Contract	

Figure 11-3: Continued

ENGINEERING CHANGE MANAGEMENT LOG					R-Rejected I-Implemented PI-Partially Implemented D-Deferred until after installation				
ECR No.	Date Rec'd	Priority (1,2,3)	Brief Description of Proposed Change	Under Investigation	Investigation Complete	Decision/ Change Should Be	Final Disposition	Comments

Figure 11-4: Engineering change management log

11.4.2 Engineering change management log

Once a change request has been proposed, it should be logged to enable you to track its existence and status throughout the program. This log, as shown in Figure 11-4, is your responsibility to maintain and should contain the following data, much of which can be found in the change request form and some of which is provided later in the process:

○ Change request number
○ Date received
○ Priority
○ Brief description of proposed change
○ Items affected
○ Current status of the change:
  ➪ Under investigation (date or check mark)
  ➪ Investigation complete (date or check mark)
○ Recommendation; the change should be
  ➪ Rejected (R)
  ➪ Implemented (I)
  ➪ Partially implemented (PI)
  ➪ Deferred until after certification (D)
○ Final disposition (may be different than recommendation)
  ➪ Rejected (R)
  ➪ Implemented (I)

⇨ Partially implemented (PI)
⇨ Deferred until after certification (D)

○ Comments

The engineering change management log is also a convenient place to track problems as well as changes. Although problems and changes are treated differently, both represent additional work. In any case, the log, if used in this manner, enables team members to check the status of both changes and problems quickly.

The engineering change management log can provide other benefits as well. For example, it is an excellent mechanism for communicating the status of changes to the customer, to executives, and to partners. Even more important, it can yield valuable quality assurance information after the program has been completed. By analyzing the number, type, source, and timing of engineering change requests, valuable insight can be attained as to how to improve on future programs. If, for example, you have received numerous requests for changes to the requirements after the final design freeze, your program may be suffering one or more of the following difficulties:

○ Inadequate analysis during the preliminary definition phase
○ Improperly defined marketing requirements and objectives
○ Incomplete review of the requirements by the quality evaluators
○ Unexpected volatility in the customer/partner's business

11.4.3 Evaluation of the engineering change request

Investigating and acting upon change requests is an important element of the project and should be managed just as any other element of the project. The person performing this investigation should complete the following areas on the engineering change request form (see the sample change request form shown in Figure 11-2):

○ Investigator – the name of the person assigned to investigate the change request.
○ Date – the date on which the investigation is completed.
○ Impact – a determination of which pieces of completed work, or portions of the product, must be reviewed and/or modified if the suggested change is implemented. All affected work products should be specifically listed (i.e., layouts, performance data, manufacturing process specifications, etc.). Any deliverables that must be revised and reissued should be noted on the checklist located on the back of the form. It is especially important to note any changes to be made to the contract, as a result of implementing a change request.

- ○ Estimated hours – an estimate of the work effort necessary to complete the change.
- ○ Estimated cost – the effect of the change on program cost, including the cost of staff hours, equipment, and supplies.
- ○ Impact on schedule – the change anticipated in the overall program schedule, as a result of implementing the requested change (including, where applicable, start and completion dates).

11.4.4 Disposition of the engineering change request

The next step in the process is to approve and prioritize the request. We recommend establishing the specific mechanism for the disposition of change requests during program planning.

The impact on the schedule and budget of the requested change is the major factor in determining the level of management needed to review a change request. Minor changes, for example, can be reviewed by the customer/partner product director, who has the authority to make decisions regarding the disposition of such changes. But the program sponsor should review major changes.

Determinations of what constitutes "major" or "minor" changes may vary from program to program. For instance, a change requiring twenty hours of work would be considered minor on a five-thousand-hour program, but major on a two-hundred-hour program. The product manager should determine what constitutes a major or minor change. All changes to the program's budget and schedule, however, must be communicated to the customer/partner, regardless of the impact of the change.

A mechanism for prioritizing changes will help in the case where a large number threaten a program's budget. On a small program, the client product manager can make priority decisions. For prioritizing changes in a large, complex program, however, it may be necessary to establish a formal change management body. Ideally, this group would be composed of the program sponsor, key partner personnel, and individuals in the customer organization who have decision-making authority.

11.5 PRODUCT DATA MANAGEMENT

11.5.1 Introduction to PDM

The engineering change process already described has the potential to generate a lot of paper, a lot of what we call product data. When combined with automated engineering design software such as CAD and CAE, the potential to become swamped in product data becomes a real risk. These factors have pushed current methods of managing and using this information to the limit. To further complicate matters, these data can come in many forms, across a wide variety of systems. Many modern engineering organizations even today will store data on a mix of computer files, paper, and aperture cards.

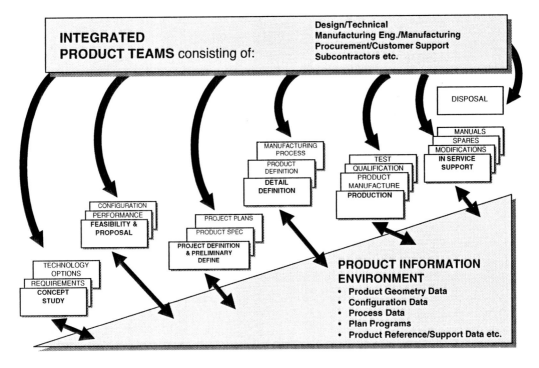

Figure 11-5: Product information environment

Product data management is the tool that helps engineers and the rest of the enterprise to manage both the product data and the product development process. In effect, the discipline of PDM will help you implement and manage the key elements of product development improvement as we have described them in this book.

As companies develop collaborative engineering practices on joint projects, the information infrastructure must support the intercompany exchange of product data along a process that extends from the customer, through the primary designer, to partners and suppliers. PDM systems integrate and facilitate the management of processes, applications, and product information that defines the product. Managing these processes and data will involve multiple systems on a variety of media.

The thesis of this book is that to truly make quantum improvements in the product development process, the practices of many distinct bodies of knowledge must be integrated. For IPTs to really be successful in implementing Concurrent Engineering, the humans must be brought together to form a cohesive team, but their data must also be brought together to permit a cohesive process. Figure 11-5 illustrates product information increases across the process life cycle.

There are several levels of product data management. At the lowest levels, electronic data "vaults" secure CAD drawings and make them available to users with check in–check out procedures to control revision status (version control). As we move up in functionality, we find workflow functionality that will notify people when an action, such as the review or approval of a drawing, is required on

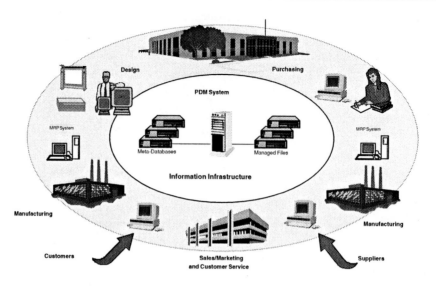

Figure 11-6: PDM concept overview presentation

a document. A defined product structure provides a part tree for relating objects in an assembly. Upper level PDM systems include methods for classification and retrieval of part information and program management functions, as described in this book. Figure 11-6 illustrates the scope of PDM system.

11.5.2 PDM functionality

PDM systems and methods provide a structure to manage and control the information used to define, manufacture, and support a product in service. A PDM system might typically contain

- ○ Product configurations
- ○ Part definitions and other design data
- ○ Specifications
- ○ CAD drawings
- ○ Geometric models
- ○ Images (scanned drawings, photographs, etc)
- ○ Engineering analysis models and results
- ○ Test procedures and results
- ○ Manufacturing process plans and routings
- ○ Assembly books
- ○ NC programs for parts
- ○ Correspondence
- ○ Audio/video annotations

Clearly this comprehensive list covers virtually all of a product's related data. A PDM system allows current data to be accessed by any authorized person within

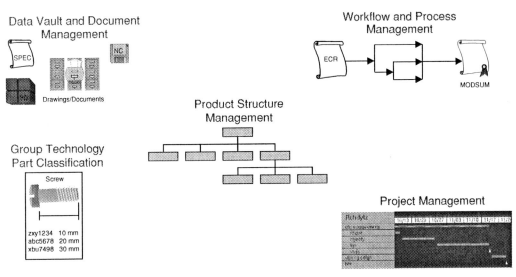

Figure 11-7: PDM functions

the enterprise. In this context, the enterprise could even be a virtual enterprise that includes partners connected by an intranet. PDM can be used to manage the product data and also to enforce discipline to the product development process. Workflow management describes a standard routing or standard process that a piece of product data would follow on the way to becoming finalized and approved. PDM systems typically contain workflow management modules that allow the modified process model to be hard-coded. By providing data management and security, PDM systems ensure that users can always share the most recent, approved information.

Linking product data through a database is a major factor in ensuring its integrity. Knowing and managing who is using the data and how provides the foundation for maintaining product integrity. Data inconsistencies can be avoided, and relationships among data can be maintained.

11.5.3 PDM functions and features

A PDM system is a software application in its own right. It requires basic functionality with which to support the design process. Figure 11-7 shows the functional view of a PDM system.

Data vault and document management

One of the key challenges in any engineering and manufacturing environment is maintaining control of design data. Different organizations within a company all have their own pieces of data related to the same product. The key to leveraging this data is making the right data available automatically to those who need it. PDM systems assist with this challenge by keeping a single master copy of the data in a secure "vault" where all changes to it can be monitored, controlled, and recorded, ensuring the integrity of the data at the time it is required. For this reason, data vault and document management is the core of a PDM system.

Reference copies are duplicate copies of the master data that are distributed to various departments for analysis, approval, or information extraction. When a change is made to data, a modified copy of the data with time/date stamps and other attribute information is stored in the vault alongside the old data. The old data remains in its original form as a permanent record.

Most manufacturing companies have developed good systems for managing the component and assembly drawings that represent their products. Unfortunately, though, auxiliary documentation about these product designs such as calculations, finite element analysis results, and NC part programs is not as systematically managed. This lack of organization means that engineers often have a difficult time accessing the auxiliary design information when changes to the design are necessary or problems arise during assembly or installation. Data management systems can manage both product attribute information and actual documents containing product data as well as the relationships between them. Additionally, this information will be readily accessible to anyone with a legitimate need to use it.

Check-in and check-out functions illustrated in Figure 11-8 are critical pieces of a PDM system for management of the data contained in the PDM vault. Check-in is the mechanism for placing data or documents under PDM control. This can happen whenever a product design element is created, updated, or approved. PDM security, including access and change control, is in force once the document or data are checked in.

Check-out is necessary when a user needs a document or data element that has already been placed under PDM control. Checking out data or documents may be required for a variety of reasons, including using without changing, modifying or updating, viewing, or marking up. When a change or modification to data is requested, the check-out mechanism triggers the PDM system to indicate that a change is in progress and may lock the information in such a way that modification cannot be made to the data by two users simultaneously.

Other basic PDM operations that fall within the data vault and document management functionality include

○ *Load*

The process of populating the PDM vault with legacy or existing CAD/CAM files and associated information.

○ *Copy*

The duplication of an existing PDM object to a new name within the PDM system. Issues regarding how dependencies are handled – whether they are automatically copied and updated – are different from one PDM system to another.

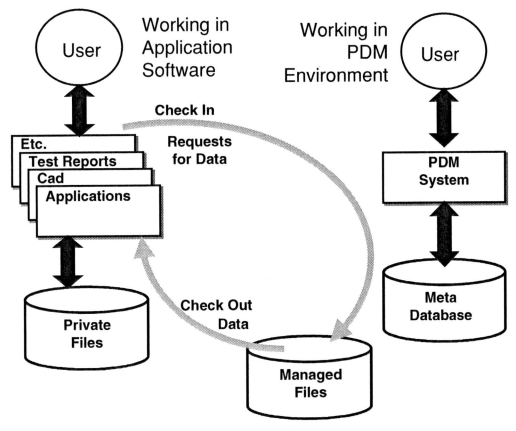

Note: A Meta Database contains information about the data (e.g., Product Structure relationship).

Figure 11-8: Check in–check out of a PDM system

○ *Delete*

The removal of PDM objects, including related attributes and metadata (data that describes other data) and associated files.

Workflow and process management

Process management is about controlling the way people create and modify data. It addresses the impact of various tasks on data, the flow of data between people, and the tracking of all events that happen during the history of a project. Business processes are made up of sequential and parallel steps, decision points, and iterations. Workflow will move the documents through the approved process.

Notifying users of work to be done and changing ownership of an object from one person/group to another during different project phases is fundamental to the workflow aspect of a PDM system. Tying the automatic processing together to support the preceding concepts and driving the detail elements of the workflow encompass the process management portion of a PDM system.

During the design process, an engineer may change a solid model numerous times, sometimes to simply explore a particular approach or test out a newly introduced concept. A PDM process management system would capture all the new and changed data as it is entered, maintaining a record of which version it is, recalling a specific version on demand, and effectively keeping track of everything the engineer does. A process history feature, or audit trail, can allow for retention of significant event information for the specific aspects of the workflow. Events that can be logged include document creation, update, deletion, distribution, and release.

Often several members of the IPT may need to be involved for input, or approval, to complete the change. The PDM system allows information to move around from individual to individual or department to department electronically and in a logically organized fashion. This aspect of process management allows for informal rerouting of design data to gain input or feedback from colleagues on ideas for design approaches, alternate calculation methods, and so on. As this information is "passed around," informal commentary with electronic "sticky notes" containing remarks or instructions can be attached.

During development of a product, many thousands of part models may need to be created, modified, viewed, checked, and approved by many different people, perhaps several times over. Furthermore, each part will call for different development techniques and different types of data – solid models for some, circuit diagrams for others, and FEM data for others. Moreover, work on any aspect of a project may have an impact on other aspects of the project so that continuous cross-checks, updates, resubmissions, and rechecks need to take place. It would be very easy, in a conventional design environment, for an engineer to invest a considerable amount of effort working on a design that has already been invalidated by the work someone else has done on another part of the project. PDM systems can manage the workflow of project data to help eliminate these disparities.

Product structure/configuration management

It is also important that the relationships be maintained between assemblies and the parts and associated data that make up the assemblies. The product structure functionality of a PDM system allows for a complete bill of material, including documents and parts, either for an entire product or selected assemblies that comprise that product. Associating attributes with both individual parts and part-to-part relationships provides better product definition and what is called "configuration management." Part attributes may apply either globally to the part, only to specific revisions, or to a specific use of a part.

Functional characteristics of configuration management typically include:

○ Support for "where-used" searches – Trace part usage through a hierarchical, multilevel bill-of-material structure.
○ Revision/version control support – Manage effectivity through product version and/or date effectivity schemes. Effectivity defines when a change is to be implemented in the manufacturing process.

○ Bill-of-material creation and support for multiple bill-of-material views:
   ⇨ Define product baselines
   ⇨ Define product features, variants, and options
   ⇨ Support rules-driven configurations
   ⇨ Compare utilities to product structure data between different views, versions, or products
○ Open, easy integration with other enterprise systems, for example, CAD, MRP, and ERP for the extraction of attribute data, the creation of bills of material, and so on
○ Support of valid "substitutes" for commonly used or hard-to-find parts
○ Support of "used-on" searches – Find all the parts of all the assemblies where a given part is used to allow evaluation of the effect of a change to that part on others in the assemblies.

Classification and retrieval

With so much data being generated and subsequently managed for various products, a technique to classify this information easily needs to be established to facilitate quick retrieval. Classification and retrieval should be a fundamental capability of a PDM system. The PDM classification and retrieval function should provide for information of similar types to be grouped together in categories that make sense for the different user environments.

Components can be stored in the database, under a variety of classes that suit a company's business needs. Classes can also be grouped together under broader headings to further enhance the retrieval process. Each part can be given its own set of attributes, and classifications of parts may have additional attribute types that are the same for every part in the classification. A broad classification example would be "purchased." All parts with the "purchased" classification would have attributes like supplier name, cost, and order lead time.

Documents can be similarly classified to specify the type of document. Document classes might include vendor specification, drawing, solid model, and so on. Each document can have attributes, like author and title. Furthermore, classes of documents can have attributes as well; for example, Microsoft Word would describe the application developed in. Relationships between documents and the components they are associated with can also be maintained. So a PDM entry for a *gear* may include associations to a two-dimensional drawing, a solid model, a spreadsheet of related calculations, and attributes about the material it is made of, the weight of the part, and the number of teeth it contains, among many other options.

Attributes that allow for easy part retrieval can be created by automatically extracting design features and parameters directly from CAD models as they are checked into the PDM vault. External component libraries, available with many CAD systems, can also be cataloged in the PDM system for quick retrieval during design. Classification and retrieval can also be an essential tool when used to gather product information into standard part, component, assembly, or product line libraries to enable the access to and use of standard parts and designs. Design retrieval and component library access are key capabilities for enabling greater reuse of existing design information and increased product standardization in the creation of new products.

PDM systems vary greatly in their classification capabilities. Some allow classification to be defined and modified as necessary, whenever the demands of organization change. Others are much more limited in their capabilities and only allow classifications to be defined when the PDM database is initially set up. In the same sense, query or retrieval capabilities vary widely also. Some PDM systems limit the scope of searches in real time, allowing larger, more extensive queries in a batch mode only. This may be especially true for queries that require searches crossing several servers in a multiserver installation.

Program/project management

Most PDM systems today provide only minimal project management functionality, such as workflow activity status information and triggers for communication of late tasks. PDM systems need to have capabilities to handle resource scheduling and task breakdown with critical path analysis, even if this functionality is available through integration with industry-accepted project management packages.

Communication and notification

PDM communication services provide the capability for sending and receiving data among the different internal applications that comprise a PDM system as well as with applications that are external to the PDM system. Communication services provide capabilities for a user to compile a diverse collection of information in a virtual project folder, refer to it by a name that encompasses all the data, like Project A, and route it to other users for review, approval, and comments. This would occur when certain events take place, like design completion or design change.

Communication and notification functionality allows users to be notified when they have a task to perform (such as design review or design approval) or when a change is proposed or implemented that may have an impact on other data for work in process. Communication and notification services also provide the ability to simply send and receive messages regarding day-to-day operations. Some PDM systems provide interfaces to standard e-mail systems for providing messaging and communication, whereas other PDM systems have proprietary e-mail modules for this purpose.

Communication and notification utilities typically provide a means for creating and managing distribution lists and monitoring due dates and sending messages as due dates approach or pass. Automatic acknowledgment to the sender when messages or sets of data are received may also be provided.

Data transport

Typically, PDM solutions provide data transport functionality so that end users who access product information do not need to know where the information physically resides or what application was used to create the data. In addition, data transport utilities may include file backup and recovery and facilitate file archiving and restoration. Basically, the data transport aspect of a PDM system isolates the user from network and operating system commands. Some of the operations that make up the data transport functionality in a PDM system include moving or copying files, either on user request or automatically triggered by an event, and routing files to the appropriate data translators when necessary, prior to sending the data to a requested application.

Data translation

PDM solutions must also include data translation functionality so that various pieces of information generated by dissimilar applications will be converted for use by additional applications. This typically includes integration with one or more industry-accepted viewing tools that can receive data in a variety of native formats and translate it for viewing in a generic form. Data translation also plays an important role in transferring data between one application tool and another. For example, utilizing FEA results in a spreadsheet for evaluation may require some form of data translation, perhaps to a neutral file format. These can be potentially difficult integrations, depending on the applications that data must be accepted from and imported to.

Viewing and markup services

Integrated image/viewing service utilities provide users with the ability to view, mark up, and convert raster images. Viewing and markup capabilities should include

- Multiple standard file viewing support including DES/STEP, IGES, DXF, HPGL, and TIFF. Most provide services that are in compliance with the CALS graphic standards.
- Native CAD file viewing for popular CAD systems.
- Annotation or markup of files with some form of graphical overlay.
- Integration with third-party viewing software.

Native file viewing is preferred over neutral file viewing because of lack of consistency among vendors in their support of standard file formats and to reduce potential errors in synchronization when multiple copies of a file are created.

Markup allows individuals the opportunity to provide comments or notes to other users of a file without changing the original file itself. Reviewers of a design might mark up a drawing for suggested changes that would then be fed back to the designer for implementation. Process planners could mark up a drawing to note special machining conditions that need to be taken into consideration when the parts are being manufactured.

Administration

Administration functions within PDM systems include support for

○ Adding users
○ Defining user security levels
○ Assigning users to groups
○ Adding data about users

Additionally, some PDM systems may provide administrative functions that offer utilities for network maintenance, event logging of use transactions, and additional application security administration. In many cases, the administration aspect of a PDM system also includes housekeeping functions like backup, recovery, and archive.

11.5.4 Users of PDM

Users of a PDM system generally fall into three categories:

○ *End users*
  The people who generate or update product information; also the people who simply need to view information about a particular product or project. End users can include project cost estimators, sales and marketing personnel, design engineers, manufacturing engineers, operations engineers, managers and executives, purchasing agents, inventory clerks, process planners, production control specialists, and assembly technicians. Actually, just about anyone who has a need to access product data within a company can be an end user of the PDM system.
○ *Product, project, or departmental managers*
  The people responsible for project planning, design review, and approval to implement the workflow of the product.
○ *PDM administrators*
  The people who provide the expertise for implementing, tailoring, and maintaining the PDM system.

11.5.5 Benefits and justification of PDM systems

A variety of yardsticks can be used to measure the benefits of implementing a PDM system. That PDM provides a vital infrastructure for corporate data is a subjective argument. PDM involves a substantial investment in software, new hardware, support and maintenance, consultants, systems integration, training, management time, and potentially business disruption. There must be *compelling business reasons* to make the investment.

PDM offers more

PDM can be shown to reduce costs through better access to consistent data and faster communications and the ability to evaluate fully more options. But PDM offers more. In preparing the PDM justification, look for *more* business, *more* market share, *more* profit, *more* quality, *more* opportunities, *more* profitable contracts when superior knowledge helps you avoid the unprofitable ones.

Concurrent engineering

PDM provides team members with access to the latest version of all necessary data at all times. Inconsistent designs that lead to fewer instances of design problems that emerge after engineering release can be eliminated. This typically results in fewer engineering change orders and, consequently, more designs that are done right the first time. PDM supports concurrent task management so a released design does not sit in someone's in-basket waiting for approval. Engineering changes can be sent first to manufacturing for acceptance or rejection of the change (communication done via e-mail). Overall, reduced time to market is achieved.

Improved design productivity

A PDM system can significantly increase the productivity of engineers and others by providing them with the correct tools to access data efficiently and without knowledge of the origin or current location of the data. Historically, design engineers would spend up to 80 percent of their time simply handling information: looking for it, retrieving it, waiting for copies of drawings, archiving older data, and the like. PDM removes this dead time almost entirely. The designer no longer needs to know where to look for data – it is all there on demand. One company conducted studies that measured the time engineers spent looking for information, filing and retrieving drawings, and making changes. They then converted these man-hours to dollar savings using an 80 percent efficiency rate for turning the wasted time into productive work. With this, they calculated that their PDM system would pay for itself in two years.

PDM offers improved use of standard parts. It also supports information exchange with partners and suppliers to provide more up-to-the minute data about parts and assemblies being supplied by outside vendors. Identification, reuse, and modification of existing designs that are similar to new ones can be done instead of "reinventing the wheel" much of the time. Rationalizing parts and encouraging the use of "building blocks" will generate significant savings. The elimination of paper files, documents that can be signed off via e-mail, and easily accessible reference and project documentation all contribute to improved productivity in the design phase of product development.

Better management of engineering change

Iterations of a design can be created without concern that previous versions will be lost or accidentally erased; PDM systems allow you to create and maintain multiple revisions and versions of any design in the database. All versions and revisions are tagged with the creator and date so that there is a complete audit trail of changes and no ambiguity about current designs. In many cases, savings from PDM system implementation have been shown through both reduction in the number of changes and the time needed to respond to a request for change.

On collaborative projects, subcontractors can be linked into the PDM system so that they too can have access to the latest product changes and specifications as they are completed.

A major step toward quality management

By implementing a coherent set of audited controls, checks, change management processes, and defined responsibilities that make up the product development cycle, a PDM system can make a big step toward establishing an environment for ISO 9000 compliance.

People, process, technology balance

PDM goes much deeper than just being a change in data, configuration, and process management; it is a change in business culture, namely how a company manages and uses its most vital resource – information. Implementing PDM can be more of a cultural than an equipment/tool change because legacy systems can still be used and integrated. The costs involved in implementing PDM must include retraining personnel and promoting the business process changes in addition to installing and integrating the software and hardware. Many people think that they do not need more technology, rather simply a way to make the technology they already have work together more effectively. A PDM system can be the tool to make this dream a reality.

Realistic implementation cost

Many PDM implementations require extensive upgrades to a company's information systems infrastructure, including new PCs, workstations, networks, and peripherals. In addition, there will be PDM client licenses, PDM server licenses, database expenses, software maintenance, auxiliary software, training, and implementation team costs. These all form part of the justification analysis.

Part 3

Deploying Engineering Process Management

12 Organizing for Deployment

Implementing the concepts outlined in this book is a major program that will affect and involve people from all areas of the company. The purpose of this chapter is to examine the people-oriented side of planning, organizing, developing, and implementing the changes outlined in this book. Implementing broad-based change to the engineering process represents one of the highest cost/potential payback improvement programs that can be undertaken at any company. A successful product development improvement initiative is the result of careful planning and organizing by the people responsible for deploying the new system. Most engineering executives know the theory of IPD and Concurrent Engineering, but few successfully deploy the concepts efficiently. On top of that, the company doesn't realize the commercial benefits. It is critical that sufficient planning occur before deployment.

The process of planning involves setting goals, selecting and developing methods for achieving those goals, and presenting the plan to begin its execution. These steps are discussed, along with other points necessary to begin the implementation.

12.1 INITIATIVE PROGRAM ORGANIZATION

The process of organizing consists of

○ Specifying the task to be performed
○ Breaking down of the total task into manageable segments
○ Establishing clear responsibilities
○ Selecting people to execute the task

Because success of the program is directly related to how it is managed, executive management should be involved in program organization. They should guide the development of the strategies to implement the changes at the company. Executive participation ensures that program goals properly support the overall business goals of the company and foster a coordinated effort among all departments. Furthermore, top management's participation in the program organization demonstrates their commitment.

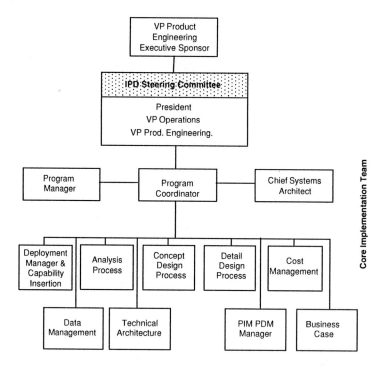

Figure 12-1: Engineering improvement initiative program organization chart

The vice president of product engineering should establish a steering committee and appoint an implementation director to direct the overall implementation and to lead the project teams who will implement the change. A typical program organization chart for an engineering improvement initiative is shown in Figure 12-1. However, the formal functional organization structure shown in Figure 12-2 cannot be ignored.

12.1.1 Steering committee

The program team reports to the steering committee to provide a vehicle for top management connection with the program. The Steering Committee should include the vice presidents of manufacturing, engineering, finance, and marketing, along with the implementation director and the program manager. The primary function of the steering committee is to provide active leadership for the program. Other responsibilities include

- Staffing the program team
- Defining the program scope
- Stimulating company commitment and dedication to the goals of the program
- Allocating resources
- Approving program budget

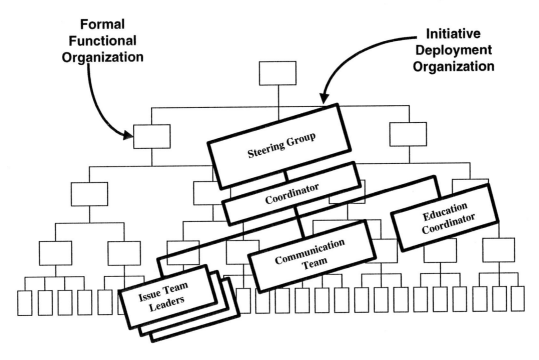

Figure 12-2: Two organizational layers – Functional and initiative

- ○ Developing overall strategies and policies
- ○ Reviewing program status on a regular basis
- ○ Attending top management education classes

The program team communicates with the steering committee to avoid surprises. Both engineering and nonengineering executives need to understand the program to ensure that an integrated effort occurs to achieve the goals. Though top management need not be involved in details, their role is to develop ideas about how the business will be run. Enlightened leadership is needed to foster full cooperation between departments. Strong management support will lead to enthusiasm for the project. The burden for stimulating top management's commitment is on the implementation director. He or she must sell the initiative to management in business terms, not through technical jargon. Outside consultants and courses can be very useful in helping to make believers of top management.

12.1.2 Implementation program director assignments

The following three key elements are necessary for effective initiative program management:

- ○ A program organization with the necessary skills
- ○ A plan for implementing the program objectives

○ A manager capable of being a synergist to utilize fully the resources and capabilities available

No team can function effectively without strong leadership. Before a program manager can perform satisfactorily, he or she needs a complete understanding of the problems, accepted techniques for solving problems, and managerial skills to get the job done. Even though every department of the company will be affected by the changes, the engineering/manufacturing areas will be impacted the most. The implementation director should, ideally, be from the engineering or manufacturing management area. He or she is the person who will be responsible for making the system work after it is installed. The program manager should be a user assigned full time to the program.

Because the team will be composed of members from various departments of the company, the program manager's responsibility will be much greater than his or her authority. The program manager must be capable of balancing consideration for people with concern for goal achievement. In addition, the program manager should possess the following traits:

○ Organizer – handle several tasks simultaneously
○ Motivator – maintain user and management commitment
○ Communicator – be tactful and assertive
○ Diplomat – be politically astute and sensitive to organizational behavior
○ Educator – provide education and training to users
○ Persistent – be strong-willed
○ Experienced – know the company
○ Knowledgeable – understand the capabilities, applications, and benefits of the process

It might be a good idea to send the selected program manager to a seminar on program management. Motivation is an important part of managing the program. The program manager should provide feedback to the team on how well its goals are being achieved. Satisfaction can be realized from seeing the achievement of even small successes and through participation in decision making. Team cohesiveness improves morale. The program manager should recognize team members whenever appropriate in program reports and correspondence.

Communication is also an essential part of managing the program. Periodic status reviews and reports are necessary to inform users and management of accomplishments, problems, and upcoming steps. Program publicity can be used for stimulating interest while keeping the organization informed. Meetings are probably the major vehicles of communication for the program manager. Regardless of the mode of communication, the information transmitted should be clear, timely, and appropriate. The program manager has two basic functions – to develop a program plan and to implement that plan.

The specific responsibilities of the program manager include

- Organizing the program and coordinating the efforts of team members and other departments
- Chairing team meetings
- Providing the program schedule and ensuring compliance with schedule dates
- Monitoring and reporting program status to top management
- Obtaining policy decisions from top management
- Identifying resource requirements
- Coordinating user education and training programs
- Representing the project team to other functions
- Ensuring that problems outside the scope of the program do not distract team efforts
- Identifying critical issues and ensuring that the right people are involved

12.1.3 Team member assignment

In many cases, predicting how well a team will perform based on the characteristics of its members is possible. As in sports, there is a relationship between the amount of individual talent and team performance; teams with better athletes do better. The team should be made up of key users who will be managing operations after the systems and processes are deployed. Representatives from manufacturing, product engineering, accounting, purchasing, industrial engineering, marketing and information systems should be members of the team (or teams). People should be carefully selected because just one individual with superior or inferior characteristics can affect the performance of the entire group. Team members should be willing to participate actively in the program – not just to attend meetings. Some members should be assigned full time to the program, and others can participate part time. This scheduling will depend on the scope of the program, existing conditions, and the size of the company. Desirable qualities for team members follow:

- Positive attitude – enthusiastic and supportive
- Good conceptual abilities – capable of envisioning how new methods will work
- Good communicator – must be a "missionary" back in his or her department
- Available to contribute effort on tasks
- Dependable – accepts responsibility to complete a task
- Understanding of the fundamental bodies of knowledge or willingness and ability to learn
- Knowledge of the company – knows how to get things done

The size of any group will affect its performance; more is not necessarily better. Generally, a group of five to eight members is ideal. When the team is finally selected, educational needs should be determined, planned, and started as soon as possible.

12.2 ENGINEERING INITIATIVE PROGRAM CHARTER

A program charter is a formal document that gives the program team authority and defines its responsibilities. The charter should list general team objectives and individual team member responsibilities, including time requirements and provisions for department sign-off. Program objectives must be specific, attainable, measurable, and agreeable to those involved. In many cases, the program is best organized into several subprograms corresponding to processes of the system or modules of the software package. This decision is dependent upon the program scope, company size, and number of program team members involved. Separate subprograms may be set up for the following modules with charters for each group:

- Concept design
- Preliminary design and analysis
- Detailed design
- Bill of material
- Routings
- Costing
- Capacity planning
- Shop floor control
- Manufacturing simulation

Individual program leaders should lead these subprograms and should report to the program manager. Figure 12-3 is an example of a charter for a bill-of-material program team. The program charter should be used to communicate the team's role and the extent of its authority to every person in the organization who has any involvement with the program.

12.3 USING AN EXTERNAL CONSULTANT

If no one on the program team has had previous experience successfully deploying an IPT and engineering process management techniques, a qualified consultant should be used. The advantages of using a consultant would be to

- Perform an objective business analysis to determine system requirements
- Plan and assist with the educational program

Bill of Material Project

❏ Define system objectives and requirements.
❏ Develop and implement an automated bill of material.
❏ Provide education and training for all users affected by the system.
❏ Disband when the system is running to the steering committee's satisfaction.
❏ Ensure metaphase BOM integrates with SAP BOM

Individual Member Responsibilities

Design Engineering Representative

❏ Establish and maintain bill of material structure.
❏ Devise engineering change control system.
❏ Ensure 99% bill of material accuracy.
❏ Administer part numbering and revision level assignment.
❏ Coordinate user accessibility of bill of material data.
❏ Define engineering department data requirements.
❏ Responsible for all design engineering facets of the program.
❏ Coordinate user sign-off approvals of system designs and specifications.
❏ Approximately 20 hours per week will be required to complete the program.

Figure 12-3: Bill of material project charter

○ Guide the process of program planning
○ Review the program plan to ensure that it is complete and realistic
○ Provide professional advice and guidance to prevent expensive mistakes and avoid major problems
○ Act as a catalyst to get things started
○ Prevent the team from "reinventing the wheel"

The consultant should not be the decision maker; he or she should counsel and coach those who make the decision. Also, the consultant should not be the program manager. The consultant should, however, aid in the initial start-up plans and then visit the team periodically to review program status. A consultant can deal quite effectively with top management problems. There is a strange phenomenon: an outsider adds credibility to the program team recommendations.

12.4 PROGRAM PLANNING

The program team has the responsibility for developing an overall plan that identifies all major milestones, along with start and completion dates. It is essential that key users and all team members be involved in making the plan. This effort will ensure agreement and enhance the practicality of the plan.

12.5 OBJECTIVES

After the program team is organized and education is under way, the first task is to define the scope of the program. Although this sounds very basic, it might seem that everyone in the organization has a slightly different understanding of the objectives and boundaries. When defining the program scope, good judgment should be exercised to limit the program to real business needs and to justify management priority. The program team must resist the temptation to include unnecessary luxuries. As soon as the program scope is agreed upon, general system requirements should be defined. A brief survey of existing systems, problems, and requirements should be documented and approved by the users. This phase of the program is very critical because it will be the foundation upon which the system is built. The future of the systems and processes depends on how well the system requirements are accepted. For this reason the system requirements document should be developed, and key users should be approved. There is often a tendency to assign a due date to the program prematurely. Defining program scope and objectives is prerequisite to preparing a program plan.

12.6 GENERAL PROGRAM PLAN

The purpose of the general plan is to show major program phases, phased in months. Management will use this plan for planning manpower and measuring progress, and the program team will use it for determining detail schedules. Just as in the reengineered Define process, phases should be broken down so that small achievements can be built upon to create a sense of accomplishment and momentum. Putting a general plan together is an iterative process – it will often change when the details are laid out.

The program plan should be completely reviewed at each phase of the program. As the program progresses, the plan must be updated to reflect any significant changes, using sound project management techniques.

Detailed task scheduling

From the general plan, a detailed weekly schedule should be made for each task, indicating who will perform each task, required labor-days, and cost, if appropriate. Task packages should be given to each subteam lead. All team members should develop such a schedule. The detailed schedule is a priority planning tool used to sequence and measure task completion. It may be wise to secure management approval of the general program plan before starting the detailed schedule.

Managing the implementation plan professionally sends a strong signal to the organization about how future design projects should be managed.

The development of a workable program plan is a necessary prerequisite to effective program planning and control.

Training

Education and training are usually the most critical tasks when deploying significant business changes. When planning the education program, the organization's needs must be assessed, its teaching methods must be defined, and a comprehensive curriculum must be prepared. The best method to assess the organization's needs is to perform an organizational readiness review.

What distinguishes process training from education in general is not just the underlying principles but the content and relationship to the desired future state. The training efforts should strive to

- Instill a product and team orientation for working in an IPT environment
- Enhance technical skills needed for the design process in a concurrent environment
- Improve problem-solving skills for cross-functional involvement within IPTs
- Clarify roles and responsibilities in a team environment (e.g., define empowerment)

A good training program typically involves interaction between an individual with training and subject-matter expertise and a trainee with specific needs, understanding, and capabilities. Both what is offered and what is heard will be affected by the individuals involved in the training exercise. It is unreasonable, however, to expect that one trainee will bring the characteristics of the entire team into that interaction. Furthermore, individuals cannot in one class take in all the knowledge presented; instead, they bring back only part of that knowledge. Therefore, a good and complete training process should include training in which individuals participate and interact as teams with the individual with training and subject-matter expertise.

By taking classes as a team, the teams gain first-hand exposure to a teaming environment. A common base of understanding is established, and skills of effective communication and teamwork are practiced simultaneously, resulting in a synergy between the various aspects of the learning to be acquired.

Training may be viewed in four parts – program specific, team building, roles, and responsibilities. The program-specific training should ensure that everyone on the program team has a common vision and understanding of the customer's needs and requirements as well as the program's purpose and product focus. The training should be an overview along with an introduction to the tools and techniques used to implement this management philosophy. Team-building training should be conducted to bring the organization together as a whole and to facilitate the cultural changes often associated with improvement initiatives. The roles and responsibilities training should address program responsibilities as well as supervisory responsibilities.

12.7 BUDGETING FOR THE ENGINEERING INITIATIVE

A budget is a formal method of evaluating means and objectives. Budgets are a guide in comparing performance to those expectations. Justification of any program should be based on costs incurred in relation to potential benefits. The program budget should initially be a rough "ballpark" estimate, followed by a firm commitment. A detailed cost-benefit analysis showing each expenditure and benefit, including information sources and cash flow schedules of expenses vs. benefits by quarter, will arouse top management attention and justify the investment.

A separate dollar amount for contingency costs should be added to the budget to provide for unforeseen expenses and cost overruns. This figure is typically between 10 and 15 percent of the costs that are not firm. The program budget may be more meaningful if correlated to specific stages of the program. For example, the purchase of hardware, software, or educational material may dictate selection of budget time periods of unequal length.

The way a budget is developed and used can have an effect on the behavior of people. The program manager and team members should participate in the budget preparation so that it is perceived as achievable. A budget is often not acceptable to management the first time around. The budgeting process is repetitive and can help identify conflicts among management.

12.8 SAMPLE DEPLOYMENT PLAN

Plans for implementation must address common obstacles to implementing reengineered processes. Figure 12-4 outlines the obstacles to successful business process reengineering (BPR) initiatives. This information is based on surveys of numerous BPR initiatives across industry.

Most business improvement initiatives follow a standard phased approach. Seven phases are identified and illustrated in Figure 12-5.

The company must commit sufficient internal resources to ensure a successful transformation to an engineering process-oriented environment. Figure 12-6 illustrates the people requirements to develop and deploy the new system for each phase.

Deployment team members must exhibit specific skill sets. Figure 12-7 illustrates the types of skills required from the program team. For example, cross-functional knowledge is required for most positions.

Obstacles to Successful BPR Implementation	
Resistance to change	60%
Limitations of existing systems	40%
Lack of executive consensus	39%
Lack of senior executive champions	38%
Unrealistic expectations	27%
Lack of cross-functional project team	26%
Inadequate team skills	25%
I.S. staff involved too late	17%
Project scope too narrow	12%
Source: Deloitte & Touche	

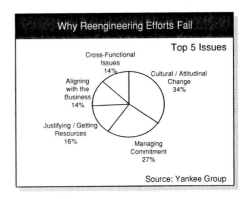

Why Reengineering Efforts Fail

Top 5 Issues

Cross-Functional Issues 14%
Cultural / Attitudinal Change 34%
Aligning with the Business 14%
Justifying / Getting Resources 16%
Managing Commitment 27%

Source: Yankee Group

"This year, American companies will spend an estimated $32 billion on business re-engineering projects. Nearly two-thirds of those efforts will fail."

Source: Michael Hammer

Figure 12-4: BPR initiatives obstacles

12.9 ROLES AND RESPONSIBILITIES OF PROJECT TEAM MEMBERS

In this section, we outline the various roles on the program team that will deploy the new process.

Subproject leads

Role
- ○ Direct and manage resources to ensure that milestones are achieved by the agreed-upon deadlines.

Activities and tasks
- ○ Report status and progress to the implementation director.
- ○ Attend steering committee meetings as required to provide specific information on status and progress.
- ○ Monitor and control budgets.
- ○ Create, maintain, and distribute the detailed workplans of the phases, including milestones and task lists.
- ○ Manage the activities of core team to ensure that appropriate work is carried out on time and within budget.
- ○ Review progress and status with team managers on a frequent basis.
- ○ Resolve issues and conflicts as they arise.
- ○ Provide deployment guidance.

Phase	Description	Major Activities
I	Project Organization & Planning	▪ Conduct project management workshop. ▪ Develop implementation strategic plan. ▪ Develop project task list and schedule. ▪ Identify roles / responsibilities / resource requirements. ▪ Establish implementation performance measures.
II	Definition of Business/Functional Requirements	▪ Review & validate detailed to-be design. ▪ Map tools to as-is processes. ▪ Document key reports & form requirements. ▪ Document interface and integration needs. ▪ Identify data volumes & operating statistics. ▪ Prepare functional requirements document.
III	Technology Evaluation & Selection	▪ Conduct technology environmental assessment. ▪ Establish selection criteria / prescreen vendors. ▪ Develop / issue RFQ. ▪ Develop / execute formal evaluation process. ▪ Select vendor(s).
IV	Process & Organizational Design	▪ Document detailed activity flowcharts / map to information flow diagrams. ▪ Conduct gap analysis for each define process. ▪ Develop final organizational structure with reporting relationships, accountabilities, job descriptions, etc. ▪ Identify and prioritize process / organizational gaps. ▪ Identify and include customer feedback and comments.
V	Change Management Planning & Development	▪ Perform change readiness assessment. ▪ Conduct core & advisory team training. ▪ Assess educational needs across all core & programs. ▪ Develop program-specific training plans. ▪ Execute training program. ▪ Develop communication plan. ▪ Document roles / responsibilities. ▪ Document performance criteria. ▪ Develop management handbook. ▪ Coordinate improvement team activities.
VI	Business Simulation/Solutions Demo Laboratory	▪ Develop business simulation plan. ▪ Develop business simulation test scripts. ▪ Identify test scenarios / build test databases. ▪ Conduct simulation. ▪ Identify / document / prioritize gaps. ▪ Develop / implement modifications.
VII	Deployment	▪ Develop / implement system interfaces. ▪ Develop conversion plan. ▪ Develop user management plan & disaster recovery plan. ▪ Perform production pilot. ▪ Perform implementation readiness review. ▪ Execute conversion / start-up for Build 1 applications. ▪ Execute conversion / start-up for Build 2 applications. ▪ Execute conversion / start-up for Build 3 applications. ▪ Execute conversion / start-up for Build 4 applications. ▪ Terminate old system applications. ▪ Perform postimplementation audit.

Figure 12-5: Seven integrated phases to the program plan

Technical team managers

Role

○ Direct and manage technical resources to ensure that technical plan milestones are achieved by the agreed-upon deadlines.

Stage	Description	Internal Resource Needs			Duration
		Leader	Core Team	Additional Subject Matter Experts	
I	Project Organization & Planning	• Full time	• Full time	• Subject matter experts (5-10%)	• 15 mths
II	Definition of Business / Functional Requirements	• Full time	• Full time	• 10 MIS (full time)	• 5 mths
III	Technology Evaluation & Selection	• Full time	• Full time	• 5 MIS (full time) • 5 IPD technical team member (50%)	• 4 mths • 4 mths
IV	Process & Organizational Design	• Full time	• Full time	• 4 program / core leaders (25-50%) • HR representative (25-50%)	• 10 mths • 10 mths
V	Change Management Planning & Development	• Full time	• Full time	• HR representative / corporate communications (25-50%) • 1-2 trainers (full time)	• 12 mths • 3-4 mths
VI	Business Simulation/ Solutions Demo Laboratory	• Full time	• Full time	• Director MIS (25%) • 1 MIS (full time)	• 6 mths • 12 mths
VII	Deployment	• Full time	• Full time	• Director MIS (25%) • Improvement team (full time)	• 15 mths • 10 mths

• Leader—strong program manager with critical skillsets
• Core team—(5-8) representing PE core, programs, operations

Figure 12-6: Internal resource needs

Skillsets/Position	Project Leader	MIS	Admin	Engineering	Operations	ERP
Teamwork	High	High	High	High	High	High
Project Management	High	Medium	Low	Low	Low	Low
Core-Functional Knowledge	Low	High	High	High	High	High
Program Leader Influence	High	Low	Medium	Low	Medium	Low
Industry Knowledge	High	Low	High	Medium	High	Medium
Training Skills	Medium	Low	Medium	Medium	Low	Low
Systems Knowledge	Low	High	Medium	Low	Low	Low
Process Understanding/Thinking	High	High	High	High	High	High
Peer Credibility	High	High	High	High	High	High
Salesmanship	High	Medium	Medium	Medium	Medium	Medium

Significance of skillset on ability to perform required task: ● High ◑ Medium ○ Low

Figure 12-7: Deployment team skill requirements

Activities and tasks

○ Create and maintain the detailed technical workplan, including milestones and task lists.

○ Manage the activities of the technical team to ensure that appropriate work is carried out on time and within budget.

○ Recommend and implement all technology solutions (including software and hardware).

○ Report status and progress to the project leaders.

○ Attend progress meetings to provide specific information on status and progress of the technical team.

○ Resolve issues and conflicts as they arise.

Process team managers

Role

○ Direct and manage technical resources to ensure that process plan, organizational plan, and change management plan milestones are achieved by the agreed-upon deadlines.

Activities and tasks

○ Create and maintain the detailed process, organizational, and change management workplans, including milestones and task lists.

○ Manage the activities of the process team to ensure that appropriate work is carried out on time and within budget.

○ Recommend and implement all process, organizational, and change management solutions.

○ Report status and progress to the project leaders.

○ Attend progress meetings to provide specific information on status and progress of the process team.

○ Resolve issues and conflicts as they arise.

Core team members

Role

○ Carry out all assigned tasks to ensure that planned milestones are achieved on time and within budget.

Activities and tasks

○ Perform project implementation activities for the technical team as follows:

　▷ Contribute toward the detailed technical workplan.

　▷ Make recommendations concerning software/hardware selection and implementation.

　▷ Assess and recommend solutions for all technical issues.

○ Perform project implementation activities for the process team as follows:
  ⇨ Contribute toward the development of the detailed process, organizational, and change management workplans.
  ⇨ Assess and recommend solutions for all process, organizational, and change management issues.

12.10 PRESENTING THE PROGRAM PLAN

To achieve full benefit from the program planning phase, the program plan must be made known to all concerned with or affected by the plan. There are two phases in presenting the program plan: the first presentation is to top management for approval, and the second is a general publishing to the company.

The implementation program director should be prepared to defend the cost, schedule, resources, scope, and benefits of the implementation. Executives will be concerned with the potential impact on the organization, especially its people, processes, and systems. Considering these dimensions in advance will make the presentation process considerably more enjoyable. Senior managers are also concerned with the support base within the organization for the project plan. For example, being able to assure the vice president of finance that the director of finance has reviewed the budget in detail is an excellent way to build credibility in the presentation.

An excellent way to kick off a program is through a letter to all employees from the president or general manager. This letter could outline briefly the purpose of the program, what the vision is, why it is needed, who will be developing it, and how it will benefit the entire organization. This letter can be followed by sending a copy of the general program plan to each department manager. Included with the program plan should be a brief explanation of each major phase. After the program plan is distributed and approved, the actual systems development and implementation phases can begin. It is now the program team's responsibility to see that the program is executed as planned.

12.11 SPECIAL CONSIDERATIONS FOR DIVISIONAL AND CORPORATE INITIATIVES

Implementing changes to the engineering process in a multicompany environment depends on the availability of engineering resources, expertise, corporate staff vs. individual company users, and the political and organizational environment of the overall company.

If the organizational structure of the company is centralized, the corporate staff will usually be the driving force in the program. Most of the team members will be corporate people; local engineering personnel will likely have little involvement until the deployment begins. However, the team should strive for a balanced input between local divisional and central staff. Central corporate staff has the disadvantage of not being close to local plant problems and needs. This sometimes causes practical problems upon implementation.

The number and skill levels of divisional engineering personnel available to work on the program will generally determine the extent of corporate staff involvement. These factors will probably vary largely from plant to plant. A completely autonomous company must be handled differently from a highly centralized one. There is much to be gained by using standard hardware, software, education, training, policies, and procedures throughout a company.

It may be advantageous to organize the central staff as a design team and set up a deployment team at each plant. The central staff, with plant input, would do most of the preinstallation work such as software package selection, education, and procedures. The local plant implementation team could then concentrate on making the system work – process redesign, data accuracy, training, engineering organization, and so on. Communication is critical in coordinating a multiplant program. Therefore, each group must make an extra effort to keep others informed of accomplishments, issues, problems, and plans. The risk of this approach is that it could be difficult to execute the hand-over from corporate to the local team.

The coordination of the local plant users, the program team, and management in a multiplant program is difficult but essential. The plant users must be involved in the planning of the program from the start. If not, the system tends to be thought of as "their" system – not "ours." Local plant and central staff team representatives should meet regularly. A good way to initiate a cooperative working relationship is through the education program. With proper management, the roles of each group will complement each other and can produce synergistic results. The timing of deployment is influenced by local cycles such as vacation shutdowns, plant "readiness," and other factors that vary at each plant. Separate implementation plans should thus be prepared for each plant. After the first plant is installed, many tasks will not need to be repeated for the remaining plants.

12.12 PITFALLS

An engineering improvement program's success requires a concentration of efforts. Failure will result from trying to do a little of everything. It is unfair and ill advised to expect a person to handle his normal job while managing a significant business initiative. Some people must keep the business running to allow others to concentrate on developing the system. If the program manager can only spend 50 percent of his or her time on the program, the chances of program failure will be greater than 50 percent. The implementation program director should not be chosen because he or she has the most time available – the skills required cannot instantly develop. Although there is a time and a place for shortcuts, the program team must not bypass proper planning and system testing. Details of a task should not be skipped to stay on schedule. Top management involvement and leadership can easily make or break the system. Management involvement must not be so strong as to cause token compliance or resentment. Also, labor unions can cause resistance to organizational change if they are not involved. Union management relations should be planned in the beginning of the program to avoid problems later on. Lack of user involvement and leadership is a pitfall that will ensure failure.

If a major engineering business improvement initiative extends too long (over three years), the company's attention span and enthusiasm will wane. Poorly defined objectives and tasks will increase the time needed to deploy the system and processes. There can be a tendency for a program team to acquire a "religious zeal" for the new system. This fervor can create tension between believers and nonbelievers and impede the acceptance of change and objective evaluation of costs and benefits.

New work systems will require exceptions from normal policies and procedures. Bureaucratic barriers must be removed to allow the company to adjust to the new work system, especially at the beginning. Conversely, overly permissive management attitudes will also cause problems. Special cutover or start-up procedures should be planned to avoid these problems. Organizing a program team requires discipline. Meeting attendance and task completion must be taken seriously, and problem areas should be recognized and resolved quickly.

By setting a tight schedule, overzealous systems analysts will not be able to spend much time exploring unnecessary "gingerbread." Most companies cannot afford to develop a "perfect" system. However, the process framework cannot be done "quick and dirty." A balanced approach, concentrating on the 20 percent of the efforts that provide 80 percent of the benefits, is recommended to achieve a maximum return on resources. Examples of these vital efforts are data accuracy, process development, education and training, and top management support – all people-related issues. The program manager must consider these issues top priority by carefully planning and controlling efforts that support them.

12.13 CONCLUSION

By following the steps outlined for organizing and planning the program, implementation will be smoother, and chances for success will be greater. Program plans should be practical, organized, and realistic. The program team will need to concentrate on the people side of the system. People are the most important part of an engineering process-oriented environment.

Management should understand the subtle relationships that exist between organizational efficiency and employee satisfaction to bring together these various resources in a creative and effective manner. Chapter 13 discusses in more detail the people and political issues that must be addressed to ensure success.

Overcoming Resistance to Change

Managing the transition from the current to a new product development process is a difficult job. One of the most critical elements of that transition is implementing a new team-based approach. Realizing tangible benefits from a team-based approach does not come easily. Resistance to change can emerge from all areas of the organization.

This aspect of the management process has consistently been a weakness for engineering-driven companies. Engineers are notorious for resisting cultural change. Traditionally, engineers focus on scientific concepts and "things" rather than emotions and people. The transition to top management is not an easy road for the typical engineer. In this chapter, we discuss the "soft" side of management. At the heart of any change initiative is balance between business needs, individual needs, and organizational politics. With the right help, this balance can be found by a company undergoing change. The organization can be improved by "engineering" the right processes, measures, technologies, and culture. Collective behavior can be predicted by the trained mind. Politics can be engineered to allow the improvement initiative to succeed. But these skills are rarely taught in engineering or business schools.

13.1 ORGANIZATIONAL POLITICS

The approach to changing the culture will need to be outside the normal project approach for an organization. To capture the organization's attention will require choosing methods that stand in stark contrast to standard operating procedures. From the very outset, the program team must free itself from the existing culture and conceive a plan of action that starts to liberate the organization from its past.

Cultural change moves at a crawl if the existing culture gets to call the shots. Or to put another way, you'll have trouble creating a new culture if you insist on doing it in ways that are consistent with the old one. After all, not following the rules is a good way to signal that the rules are being changed. The old culture is designed to protect itself, not to bring about its own demise.

Management view of the organization

Individual view of the organization

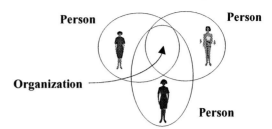

Figure 13-1: Management view versus individual view

Organizations keep falling into the trap of letting the existing culture dictate the terms and conditions of how the changes will be carried out. Instead of drawing up a course of action that is deliberately foreign to the existing culture, they're prone to adopt strategies that are too compatible with the existing culture. This makes no more sense than trying to win a war while letting the enemy design your battle plans. So why does it happen?

It happens because culture wields great power over what people consider permissible and appropriate. The embedded beliefs, values, and behavior patterns carry tremendous weight. The culture sends its energy into every corner of the organization, influencing virtually everything. If you're not careful, the old culture will permeate your game plan for change like a lot of bad wiring and will short circuit your efforts.

It just doesn't make sense to try to change culture according to the old rules. The rules themselves are part of the problem. Choose a change strategy that runs contrary to cultural habits and defy tradition. Disregard the managerial norms that safeguard the established (but outdated) way of doing business. Flout the values and symbols that are relics of an antiquated culture because your actions become symbolic. The program teams style, technique, and overall strategy because culture change should be alien to the status quo. And the team leaders must understand political management to be successful.

13.1.1 Managing organizational politics

Managing organizational change embraces the "human" aspects of the process. Change models tend to treat people as one element of a complex mechanism called an organization. This leads to the assumption that organizations, like machines, can be changed at will. In truth, it is the other way around – organizations are entirely composed of people, each of whom has a life outside of work. An organization can only change if the people who participate in its activity wish it to change. Figure 13-1 illustrates the difference between management and the individual view of the organization.

Consulting approaches have traditionally focused on defining what should change in an organization for it to achieve better economic performance. The management is then left to persuade the people within the organization to adopt the proposal. However, professional management consultants are increasingly measured not on the quality of their designs but on the impact on the business. The "rightness" of the solution is increasingly less important than whether it can be implemented.

This chapter seeks to address how to define a solution that will be accepted and how to get people within the organization to implement it. The purpose of this chapter is to introduce the concepts related to the sociopolitical management of change. This chapter provides a theoretical framework for an approach to overcome resistance to change by explicitly managing the sociopolitical aspects of the organization.

The framework (see Figure 13-1) has borrowed freely from a number of disciplines, and notes in the text refer to a number of mainly academic sources:

○ Psychoanalysis
○ Psychology
○ Mathematics
○ Military strategy
○ Behavioral science
○ Management science

There is nothing more difficult to carry out, nor more doubtful of success, nor more dangerous to handle than to initiate a new order of things. For the reformer has enemies in all who profit by the old order, and only lukewarm defenders in all who would profit by the new order.

The lukewarmness arises partly from fear of their adversaries who have the law and tradition in their favor and partly from the incredibility of mankind who do not truly believe in anything new until they have had actual experience of it.

Machiavelli (1514)

There is a sense in which the defeatist management science talk is correct (that nothing can be done about politics). The problem is this. When used in everyday language, the word "politics" usually refers to what is *never* explicit; it refers to assumed hidden agendas driven by consideration of power, which may be motivating individuals but will never be made public. In this sense, "politics" *is* unanalysable. It operates at whatever level is meta- to that at which the analysis is conducted.

Rather (than extend political analysis too far and) move in the direction of the amateur psychoanalyst, it is better for would-be improvers of human situations to assume good faith on the part of their fellow human beings, an assumption which often yields surprising dividends of goodwill.

P. B. Checkland, University of Lancaster, UK (1986)

The biggest barriers to change are human, not analytical. A recent survey of one hundred U.S. organizations that had completed or were undertaking reengineering programs identified that six out of ten critical success factors and seven out of ten barriers were related to human or political issues. Issues such as middle and line management resistance, inappropriate prevailing culture and political structure, employee fear and resistance were the most common. From the same survey, the most common lesson learned was that the participants needed "to pay much earlier, focused attention to the human and political factors inherent in reengineering." Seventy percent of Forbes 500 CEOs see employee resistance as being the number one impediment to organizational change

Academic research into the nature of change provides overwhelming empirical and theoretical evidence that social and political factors drive behavior in organizations: from psychoanalysis and the Hawthorne experiments of the 1920s and 1930s, through the Tavistock experiments of the 1970s, to more recent academic management science.

This research supports the following everyday experience:

❍ Whether we like it or not, "politics" is important in most organizations, especially at the senior level.
❍ People are enthusiastic about change if they get something out of it; otherwise, they are hostile.
❍ People are often more concerned with titles denoting power and social position than what they actually do.
❍ As individuals, we view change to our own lives primarily in terms of how it will affect us personally.

13.1.2 Focus on the future

Leaders of change initiatives should not get bogged down in the endless task of "culture analysis." Culture change should be guided by where the organization needs to go, not by where it's been. Executives don't have time to sit around and sift through the sands of our history, trying to figure out the intricate details of who they are and how they became that way. It's a seductive exercise, but it simply takes too long. Sometimes it's just a stalling tactic or a safe way to camouflage resistance to change. Nevertheless, it is, above all, an unnecessary drill.

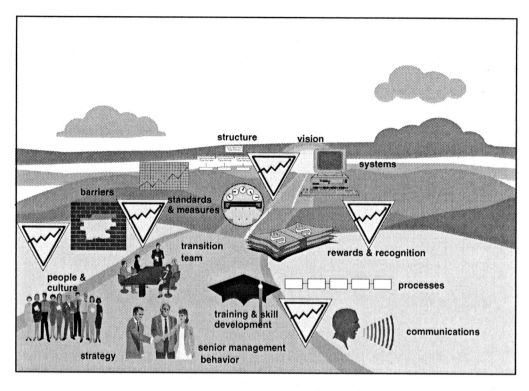

Figure 13-2: Change management guideposts

Instead of wasting precious time contemplating the company's navel, those who want to implement change must simply be clear on what the company needs to be. Analyzing the present culture is like going to history class, when you could learn more valuable lessons from studying the future. The team needs to take instruction from tomorrow – that's where they will find the answers they need.

A quick scan of the future tells us, for example, that the organization's very survival depends on speed. Competitive advantage will come from being faster than the next guy. The race will be won by the swiftest. So it is important not to burn up precious time and waste resources looking backward.

Simply by starting to move toward tomorrow, the team will know what they need to start shooting for. The window to the future gives better guidance than the mirror. This is a time for action, not introspection.

To manage the change process effectively the team must develop "guideposts" (illustrated in Figure 13-2) for each key, strategic "lever" along the way.

13.1.3 Political tactics

Most consulting approaches ignore social factors and consider it unethical to even recognize that politics within the organization exists. There is substantial material on changing business processes, but the popular, largely U.S.-orientated

If I do not see political behavior, it will not affect me.

If I do not listen, I can ignore people's views and concerns.

If I do not talk about politics or social behavior, it will not affect me.

And we wonder why change programs fail

Figure 13-3: Political management ignorance

literature on change generally ignores the importance of social interaction within organizations. Politics is never even acknowledged, although it is often referred to as "wooly" stuff or "smoke and mirrors." Figure 13-3 illustrates this ignorance. Where resistance to change is discussed, it is treated superficially and usually relies on the importance of leadership, reducing an inherently complex and paradoxical problem into simplistic models.

When change management practitioners encounter resistance to change, they will typically facilitate open communication and participation in an attempt to agree upon common goals. Individual practitioners with less experience may attempt to overcome resistance to change through open communication, team building, and consensus creation. Although this approach may be effective in many cases, it cannot deal with determined opposition where the change objectives conflict with political and social priorities. Typical behavior under these circumstances includes

- ○ Critical issues facing the organization are put off limits, making participative involvement impossible.
- ○ Issues are addressed behind closed doors through political negotiation, not analysis.
- ○ The change devolves into a "war of attrition" as individuals and groups resist the new approach.
- ○ Individuals and groups refuse to participate or, at best, participate in a superficial way.

There are two approaches to implementing change – the consultative approach and the sociologist approach. These paradigms, which are illustrated in Figure 13-4, are so strong that the two views are rarely combined. To the consultant, politics and social issues should be ignored. The link between politics and tangible results is unclear and should be avoided. To the sociologist, self-interest

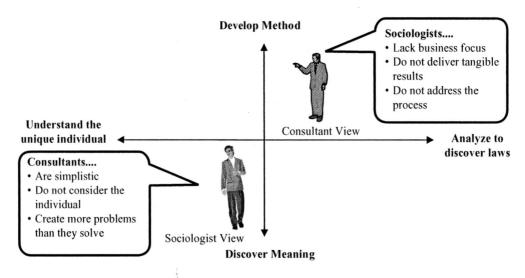

Figure 13-4: Sociologist versus consultant view

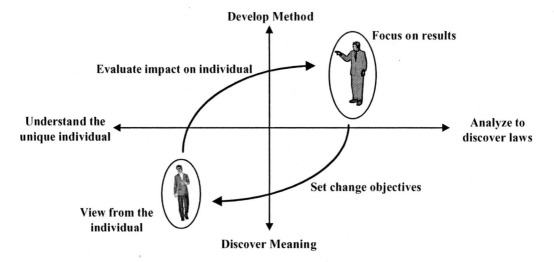

Figure 13-5: Individual view of the world

and the view of the individual is paramount and more important than economic performance. A person's view of the world should be respected above all.

Managing resistance to change effectively encompasses both paradigms. Successful change programs change the balance point as resistance develops (see Figure 13-5). This is especially true in an engineering-intensive organization. As we have already noted, engineers by training are used to focusing on things rather than people and emotions. The interests of the people in an organization undergoing change are neither "right" nor "wrong" – they are merely "compatible" with the objectives of the change program or "incompatible." If they are incompatible, they will lead to resistance. The aim of sociopolitical management is to align the

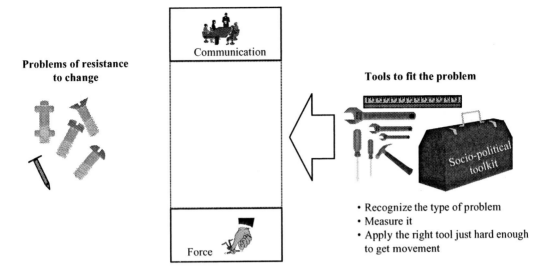

Figure 13-6: Political tools to overcome resistance

individual's agenda with the change objective. This will lead people to want to change, and implementation will be easier and quicker.

The expertise of the best change agents can be codified to provide a toolbox of approaches to overcome resistance to change. The objective of managing the political environment in organizations is to utilize these tools every day to deliver the benefits of the change, as illustrated in Figure 13-6. The best change agents intuitively know how to persuade people to change. They have many tools and tricks of the trade to achieve this. The aim of managing organizational politics is to learn the theory behind the approaches, codify them, and make them accessible to all.

Managing organizational politics is not a stand-alone "product." It addresses how resistance to change can be overcome, not what the change should be. Like project management, it will not deliver tangible results by itself, but it should be part of a successful management style. By going back to the basics of human behavior, managing organizational politics embraces all the human aspects of change. It supplements project management and participative change management approaches, as illustrated in Figure 13-7.

13.1.4 Reasons for resistance

Change programs seek to improve the performance of the economic system illustrated in Figure 13-8. The economic organization is "open" in that it performs a primary task on its environment. Skills and knowledge of people are applied to business processes through organizational systems and structure. Strategy defines the purpose, objectives, and scope of the organization and is influenced by external, internal, and historical factors. The system is relatively static and, in the best

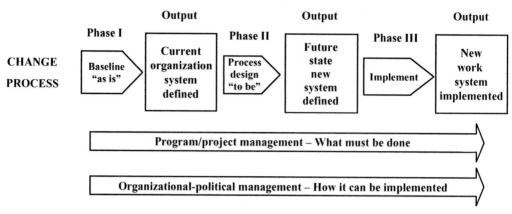

Figure 13-7: Business transformation including political management

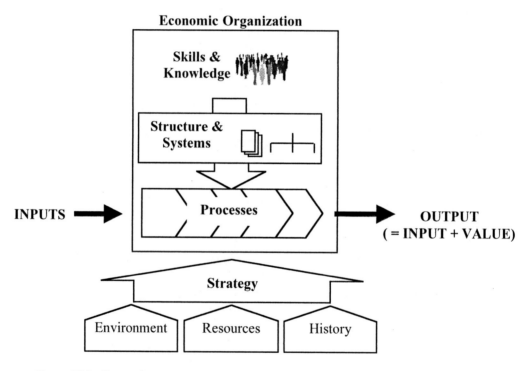

Figure 13-8: Economic system

organizations, is internally consistent and well controlled. An economic organization can be a country, a city, a business, a government, a family, or an individual.

Resistance to change comes from the economic and political "shadow" system illustrated in Figure 13-9, whose purpose is to fulfill the personal needs of employees. People come to work to have their personal needs fulfilled. The shadow system that fulfills these needs is said to be closed as it is and self-sustaining. In an engineering group, the shadow system is the bond between the engineers in the design group. The connection between degreed professional engineers is

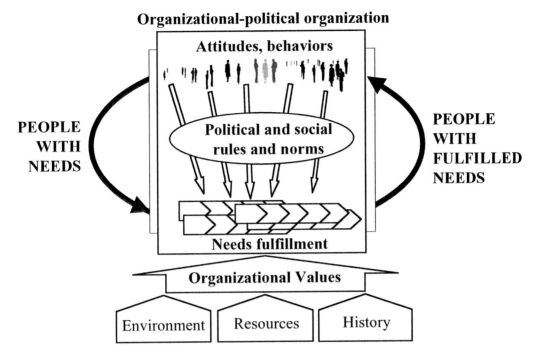

Figure 13-9: Economic and political system

often greater than with the rest of the company. This connection has deep roots established in the engineering schools at universities. And this special affinity is itself a subculture and subsystem within the company culture. This system is highly dynamic and is often fragmented, inconsistent, and in constant conflict. A "functional" system results in "normal" and healthy behavior, high morale, and satisfaction. A highly "dysfunctional" system results in destructive interactions and personal dissatisfaction.

People resist change when the change challenges fulfillment of these needs. Human behavior is driven by personal needs and is illustrated in Maslow's hierarchy in Figure 13-10. As needs for shelter and security are satisfied, the pursuit of influence over resources (political power) and growth increasingly dominate individual motivation. The different agendas of groups within organizations tend to reflect this hierarchy. Senior management tends to be very concerned with the "political" agenda of power and influence, whereas the workforce tends to react to "security" issues like job security and wages. All are strongly protective of social relationships. Reaction to change – resistance, compliance, or leadership – will be determined by perceived impact of the change on individual needs.

The type of message typically presented to different hierarchical groups during a change initiative illustrates this. Although never explicitly recognized, the main concerns of employees within organizations broadly follow this hierarchy of needs. Typical messages that can be used to make people feel comfortable with change are illustrated in Figure 13-11.

	Maslow's hierarchy of needs	Provided by organization
Self actualization & growth	Development & change
Development of ego and sense of self	Power, influence and self-esteem
Need for social interaction	Belonging to social group
Need for safety and security of position	Comfort, security of employment
Physical need for shelter and food	Reward & job survival

Last need to be satisfied ▲ **First need to be satisfied**

Figure 13-10: Maslow's hierarchy

	Prevailing need	Objective of resistance	Message
Senior management	Power and influence	Protect power	Position will still be important
Middle management	Social position	Reduce uncertainty	The new organization will be a better place
Employees	Security and shelter	Protect jobs & wages	Company success will protect jobs

Figure 13-11: Messages by hierarchy position

If the processes in the economic and shadow organizations are misaligned, employees will tend to behave against the best interests of the organization. This is illustrated in Figure 13-12. Such organizations are perceived as being "political," even though all organizational interaction is political. When the economic and needs fulfillment processes are aligned, they reinforce each other, just as improving performance of the economic task is directly linked to fulfillment of individual needs. If they are misaligned or disconnected, then people attempt to fulfill their needs in ways that detract from the primary task, waste energy, and reduce the economic organization's effectiveness.

The unwritten rules that govern behavior in the shadow organization are driven by the rules and systems of the economic organization, and vice versa. Each rule, structure, and incentive in the "economic" organization has consequences that manifest themselves in "unwritten rules" in the shadow organization. For example, "Broad experience is important" can be translated "Job hop"; "Pay is a function of resources controlled" is often perceived to mean "Increase personal empire"; and "Must be an engineer to get ahead" is believed to mean "Increase personal empire." Thus, decisions in the economic organization are constrained by political considerations from the shadow side.

Successful behavior is rewarded by possession of "commodities." People resist when change threatens to redistribute commodities or alter their value. Success in sociopolitical terms is measured by possession of scarce resources, or power

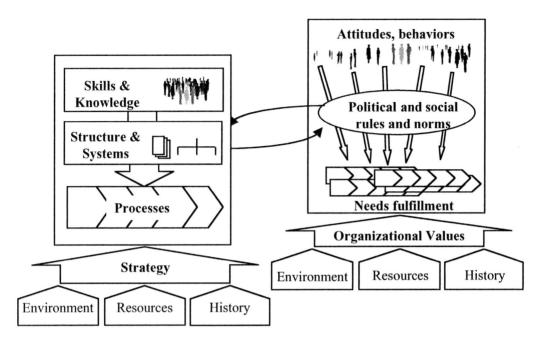

Figure 13-12: Economic and shadow organization relationship

"commodities." They are often difficult to identify, but they are the subject of most "political" conflicts. Typical commodities include

- ○ Salary
- ○ People
- ○ Budgets and spending authority
- ○ Membership in key decision-making bodies or "clubs" – the good old boys
- ○ Job titles/physical locations/symbols of authority

Change inevitably redistributes or threatens redistribution of these commodities. In political organizations, redistributing power is seen as a zero-sum game – one person's win is another's loss. This can be seen in major conflicts around the world. In Northern Ireland, the Catholics are suspicious if the Protestants are happy and vice versa. Generally, people resist loss harder than they drive for acquisition; therefore, the overall "balance of power" is generally against change. This is illustrated in Figure 13-13.

The most significant resistance to change will often come from the "brokers" who control the commodities. When fundamentally changing the product development process, often the broker is the vice president of engineering behind the development of the new processes. The most powerful people in the shadow organization are the brokers who bestow or remove commodities according to widely understood, but largely unwritten, rules. The less well defined the rules, the more

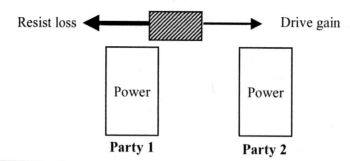

> **The more misaligned the organization, the more change is likely to redistribute commodities, and the more "political" resistance is likely.**

Figure 13-13: Zero-sum game

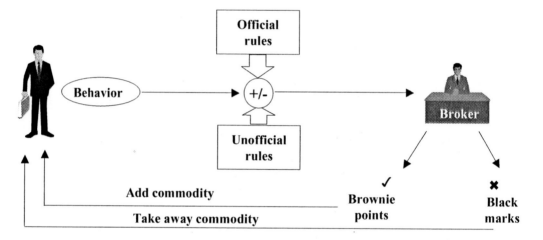

Figure 13-14: Broker influence

negotiation and favors come into play, and the more influential the brokers. This concept is illustrated in Figure 13-14.

Attempting to force change that conflicts with the unwritten rules increases resistance. The natural reaction to resistance to change is to push harder and exert more force. However, this tendency increases the threat to the shadow organization, and so increases resistance. Behavior of individuals can become increasingly irrational and desperate as they struggle to understand what is required of them. Figure 13-15 illustrates how, as pressure increases, change becomes more and more political and more difficult to deal with. A person who is actively resisting the change will conspire to sabotage the effort when the pressure is high.

In addition to such political resistance, "emotional" resistance occurs when change disturbs systems of relationships and social interaction. Individuals within organizations form social systems or "groups" to provide social interaction and to deal with the emotional pressures (anxieties) of life and work. The group develops

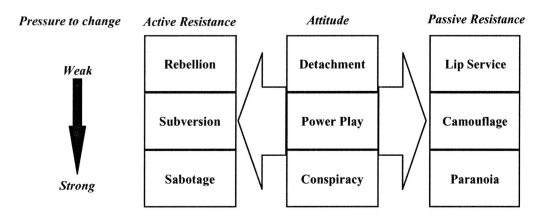

Figure 13-15: Impact on the shadow organization of changes that conflict with unwritten rules

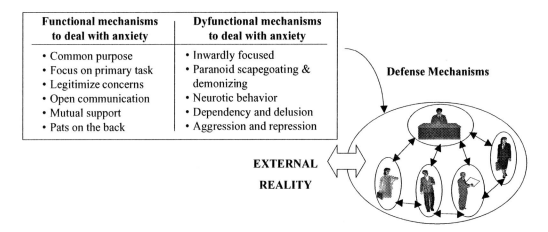

Figure 13-16: Dysfunctional social systems

norms of behavior and defense mechanisms of varying functionality. Socially functional groups provide a supportive environment able to deal constructively with anxiety. Socially dysfunctional groups become introspective and destructive and increase individual anxiety. This is illustrated in Figure 13-16.

Resistance to change will be greatest when economic and shadow organizations are misaligned or when social systems are dysfunctional, as illustrated in Figure 13-17. When primary and shadow systems are aligned and functional, change will be more acceptable. Individuals can rationalize that improved organizational performance will enhance their personal needs. Individuals are less apprehensive when they have confidence that "society" will provide support during transition. Poorly aligned organizations have much more complex reactions to change and are much more resistant to it. Political organizations resist change to power structures. Dysfunctional social structures react unpredictably and are destructive.

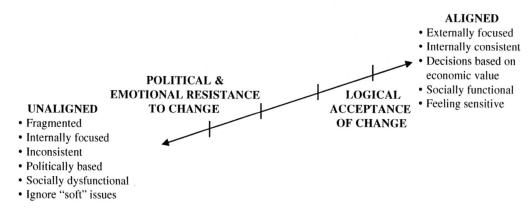

Figure 13-17: Aligned versus unaligned systems

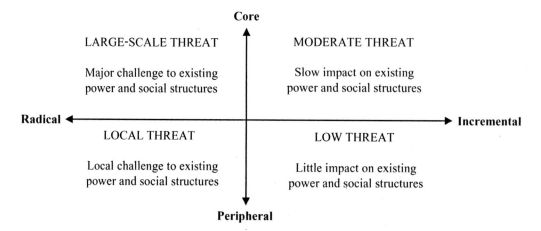

Figure 13-18: Most business transformation assignments pose a significant threat to at least part of the organization

Resistance increases with perceived threat, as illustrated in Figure 13-18. Change will be most threatening if it is "radical" and challenges the "core" of the organization. Core issues directly affect the perceived purpose or values of the organization, individual, or group. Typically, changes to processes associated with the primary task and changes that threaten central beliefs, written or unwritten, which the organization has about itself, are considered the most radical. People perceive radical changes to require major behavioral change or pose substantial risk to the performance and execution of the primary task to individual power bases or social structures.

Politically motivated resistance will be greatest when a new process threatens radical change to core values in a political organization, as illustrated in Figure 13-19. Participative and analytical approaches are almost entirely powerless in the case where opposition is politically and emotionally motivated.

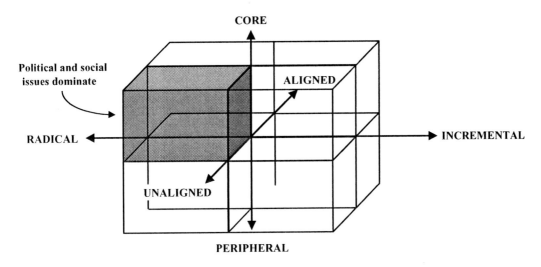

Figure 13-19: Radical change unaligned at the core

Arguments based on economic logic do not address the reasons for resistance and have little relevance to the political agenda. Participation will be blocked as key players remain uninvolved, critical issues are not addressed, or continual attempts are made to hijack the program to serve "political" needs.

13.2 ADOPTING THE TEAM-BASED APPROACH TO THE ENGINEERING ORGANIZATION

Many companies attempting to implement the concurrent engineering approach focus on technological goals. They spend millions of dollars studying the latest virtual product development tools, as though the only change is a technological one. Nothing could be farther from the truth. Just as crucial, though often overlooked, is the need for a highly interactive, multifunctional group of participants. This group includes both program and functional participants. Successfully implementing the team-based approach is central to implementing a Concurrent Engineering process.

The distinction between establishing concurrent development teams and implementing a team-based approach should not be understated. It is easy to form a collection of people and call it a team. In the team-based approach, a team is a group that, as a whole, agrees on its purpose and then works toward it. Not knowing the difference, managers might think all they need is to assemble a skillful, multifunctional group. Experience from other groups similar to IPTs such as project teams, task forces, and decision-making groups shows that it takes more than that. Before a group can become a team, it must overcome the obstacles to teamwork.

13.2.1 Obstacles to IPT effectiveness

Team building techniques have been around for years. Everyone has heard of forming, storming, norming, and so on. Applying these valid approaches is pointless unless the politics are addressed. This section describes how to address politically motivated resistance to the team-based approach.

13.2.2 Differences in orientation

Teamwork requires that team members agree on team goals, problems, and solutions. One hallmark of IPTs is diversity, and although diversity helps stimulate the exchange of issues and opinions, it does nothing to enhance agreement. People from different functional areas have different orientations and dissimilar preferences.

For example, in comparing production, marketing, and engineers and science workers, Lawrence and Lorsch found that

- ○ Production people are more concerned with getting the job done, while sales people are more concerned with maintaining coworker relationships.
- ○ Sales and production people want short-term results and rapid feedback, whereas engineers and scientists are satisfied with longer-term, "delayed-gratification" results.
- ○ Even between similar functional areas like product engineering and manufacturing engineering, attitudes, goals, and viewpoints often vary greatly.

13.2.3 Inequalities among members

In an effective team, everyone participates and believes that his or her views are welcomed and valued. Often participation is repressed by significant inequalities between members. Three sources of inequality are common in integrated product teams. Kanter described inequalities stemming from status, personal features, tenure, and activity level. When members come from different functions or levels, a status hierarchy that treats members unequally can develop. The chief executive officer (CEO) might be given more talk time and consideration than, say, a director or vice president. Similarly, design engineers might be treated with more respect than manufacturing engineers, and engineers might be given more consideration than technicians or machinists. Such differences prevent members from openly expressing ideas or disagreements. These differences can frustrate a CEO. Often a CEO in his or her desire to change the company will resort to hierarchical force, which has many downsides, as illustrated in Figure 13-20.

Participation is also affected by members' abilities to express ideas, develop arguments, or reach decisions. IPT members from marketing might be more experienced in working in a group and have better team skills than, say, engineers,

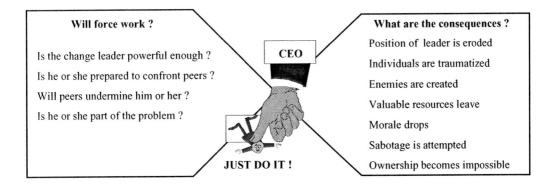

Will force work ?		What are the consequences ?

Will force work ?

Is the change leader powerful enough ?

Is he or she prepared to confront peers ?

Will peers undermine him or her ?

Is he or she part of the problem ?

CEO

JUST DO IT !

What are the consequences ?

Position of leader is eroded

Individuals are traumatized

Enemies are created

Valuable resources leave

Morale drops

Sabotage is attempted

Ownership becomes impossible

We do not have time for a war of attrition

Figure 13-20: Hierarchical force is a tough way to bring about change

who are accustomed to working alone. These differences might prevent people from speaking up or being heard.

Tenure and degree of involvement also influence participation. The process is circular; the more people participate, the more open they become, the more the team listens to them, and so on. Meanwhile, it becomes harder for newcomers and occasional members, such as suppliers and customers, to speak up or to be heard.

13.2.4 Too much spirit

Team camaraderie promotes group vitality, so it would seem to be a good thing. But it can cause groupthink – the team thinking that it is invincible to the point where it discredits even obvious signs to the contrary. Presumably, groupthink is unlikely in IPT because customers and suppliers will quickly point out the team's bad ideas. But there is no assurance these people will always speak their minds, or that their views will be given enough weight. Paradoxically, an IPT that has been very effective at tearing down walls between functions may, in effect, build a wall between itself and important outsiders.

13.2.5 The team myth

Despite differences between group members, members think they are not supposed to be different because they are on a team. Kanter calls this the team myth. When some members realize they often disagree with the group or are being treated with less respect, they think they are the "oddballs" and react, typically, by participating less. Detecting this, other members, especially the very active and involved, back off so that the others will be able to participate more. If this does not work, the result, ironically, is less participation all around.

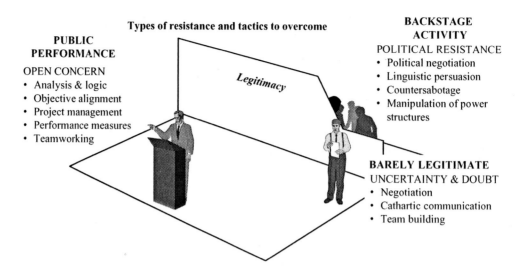

Figure 13-21: Types of resistance and tactics to overcome

13.2.6 Hidden agendas and politics

In an IPT comprised mostly of part-timers, members often still maintain primary allegiance to their functional departments. They may never actually see themselves as members of the IPT. As a result, the team will inherit interfunctional competition or rivalries that exist in the larger organization.

The openness of tactical interventions varies with the level of conflict. As the team builds consensus, tactics move from backstage into the public arena. This is illustrated in Figure 13-21. Political conflict always starts backstage, veiled by a curtain of legitimacy. Political opposition is often legitimized through apparently economic logic. Emotional resistance is even deeper in the background. The real reasons are often unknown even to those resisting. Situations of direct conflict and entrenched resistance must generally be addressed backstage. As consensus is built for change, interventions can move progressively into the public arena.

Typical counterimplementation strategies rely on increasing ambiguity. These strategies include

- ○ Diverting resources with conflicting priorities and by sharing facilities
- ○ Exploiting inertia by having endless reviews and task forces or creating dependencies on another project
- ○ Keeping goals vague and complex
- ○ Encouraging organizational ignorance and ignoring people issues, knowing that it will kill the project later
- ○ Bringing in representatives from all key stakeholders and paralyzing decision making

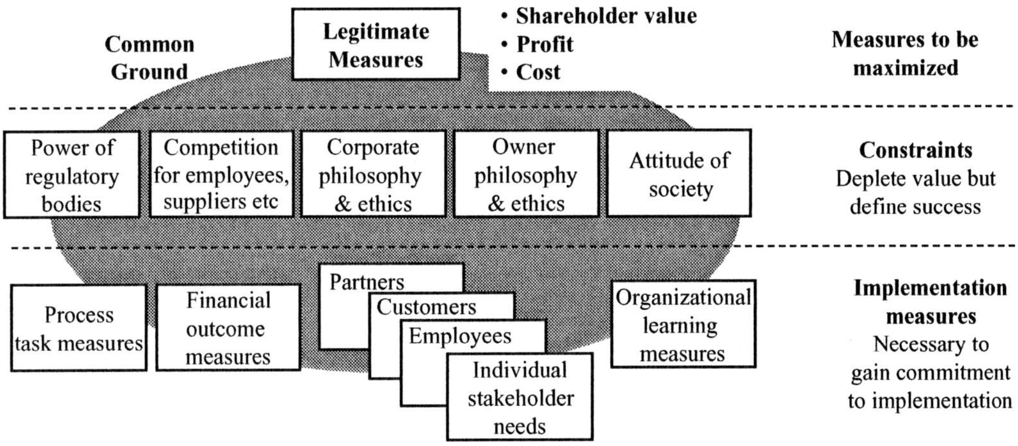

Figure 13-22: Common ground value proposition

○ Dissipating energy by surveys, reports, trips, and meetings
○ Attacking champions influence and credibility
○ Keeping a low profile, agreeing with everything, and doing nothing

13.2.7 Weak or inappropriate team goals

One problem with many teams is that they start off assuming that everyone knows and agrees to the team goals. Thus, they never question or define clear goals. The predicament is exacerbated in an IPT because, in the absence of clear consensual goals, members from different backgrounds automatically interpret the goals to suit their own functional or personal bias.

By carefully constructing benefits for key stakeholders, a "common ground value proposition" can be developed and supported by a sufficient balance of power to achieve successful implementation. This is illustrated in Figure 13-22. Most people who are required to implement change will be unconcerned by the measure of success, unless a WIFM (what's-in-it-for-me) solution that will not violate undeclared constraints is provided for all key stakeholders. The solution may deplete the value but may be broadly accepted. Any proposition that has widespread appeal is an "open door" to change and should be exploited to gain a good position within the organization whether or not it is directly related to the overall measure of success.

13.2.8 Autonomous team

An IPT cannot be a committee where members maintain outside duties. Members must be relieved of current obligations and placed in a separate unit (a program) to ensure that everyone has the same priorities and allegiances. It might seem

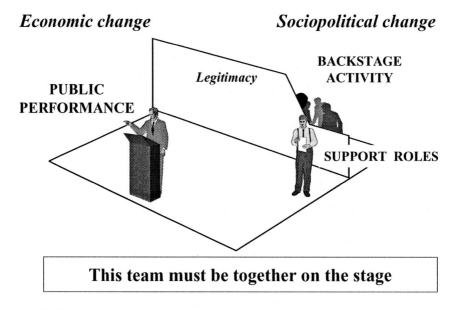

Economic change   _Sociopolitical change_

PUBLIC
PERFORMANCE

Legitimacy

BACKSTAGE
ACTIVITY

SUPPORT ROLES

This team must be together on the stage

Figure 13-23: The two change agendas must be managed in parallel

easier to keep members in home departments working on additional, non-IPT tasks. In cases where compatibility with existing processes or products is the goal, it might even be advantageous, but the result is always less commitment, communication, and coordination than with an autonomous team. There always will exist two agendas – the program and the function or the public performance and the backstage activity – as illustrated in Figure 13-23.

13.2.9   Full-time, full-duration team

The autonomous nature of the team assumes that members are committed full time for the duration of the effort. Even though each member's expertise is needed only at certain times, ongoing member involvement ensures continuous representation and full agreement with all decisions throughout the product development life cycle. That is what _concurrent_ means.

In instances where input from specialists, suppliers, or customers is insufficient to justify full-time involvement, there should be some other, well-defined connection to the team. For example, a team desk where a participant sits once a week or whenever working on the project would provide a symbolic connection. The team should make frequent scheduled visits to their facilities to keep customers and suppliers involved. Some IPT members are incorrigibly bad team players; rather than try to incorporate them into the team, it might be better to give them clear assignments and work latitude. A liaison can then meet with them periodically to review progress and give updates.

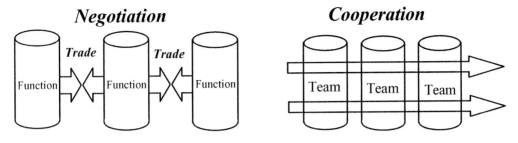

Figure 13-24: From negotiation to cooperation

13.2.10 Colocated team in a program management organization

When the team is located in one office, ideally one with no walls between desks, communication, which would otherwise take weeks, will happen immediately. Formal reviews are replaced by thousands of informal chats. Members become familiar with each other's perspectives and integrate them into their thinking, which reduces disagreement later on. This open office environment reinforces team self-identity and diminishes or erases any external sources of status differences. The goals of the program become more important than the goals of the function. This change requires that the organization move from trading to cooperation, as illustrated in Figure 13-24.

If the goal is to produce novel ideas with minimal outside influence, the team should be somewhat isolated from the main organization. Given the more common goal of developing concepts for production, it is better to locate near or on the premises so that members can check ideas with functional colleagues.

Full implementation of the concurrent engineering approach requires a fundamental cultural change to attitudes, thinking, and behaviors. Expected new behaviors include working in cross-functional teams employing horizontal communications, customer-focused thinking, and rapid decision making.

13.2.11 Small teams are more effective

Teamwork is enhanced by small size. An IPT of eight to twelve is large enough to represent all major functional areas, the customer, and suppliers, yet small enough to allow effective communication and to encourage team commitment. Though a larger size might be necessary to get every viewpoint, it risks alienating members from the team. It is better to keep the team small by including members who have multiple skills and broader outlooks. Programs that require large numbers of participants should be subdivided into smaller, strongly integrated subteams that relate to the work breakdown structure. This organization is explained in Chapter 9.

13.2.12   Team rewards

Team-oriented rewards encourage team effort. Whenever the team satisfies or exceeds requirements, the members might receive an opportunity to be on another leading-edge team (presuming the last team experience was good), write-ups and photos in magazines and sales brochures, expense paid vacations, or recognition by someone they hold in esteem. Cash, unless ample, is usually less effective as a team motivator than rewards custom-tailored to the team.

13.2.13   Team of doers

Besides making decisions, IPTs "do" most of the necessary work using their own labs and shops. Though most IPT members are specialists in their own right, each must assume a wide range of obligations. To this end, members must be "can-do" people willing to visit suppliers and customers, do drafting, modeling, light assembly work, or whatever else is needed. This, of course, increases both the team's vitality and its appeal, and once outsiders see the range of the team's accomplishments, lines will form to volunteer for the next IPT effort.

13.3   TEAM LEADER CHANGE MANAGEMENT SKILLS

Though a high-performing IPT may appear to work well without a leader, usually in the background there is someone very busy leading it. What makes a good IPT leader? Frequently we focus on three qualities – technical knowledge, management, and ability to help – but they alone are not enough. Those who influence the decision makers and implementers are also key stakeholders. An organization is made up of informal networks, as illustrated in Figure 13-25.

13.3.1   Clarify and build commitment to the team purpose

The research on high-performing teams indicates one constant – the leader's ability to clarify and convey the team's purpose and objectives. Vaill, who has studied hundreds of high-performing teams, calls it purposing. Good leaders create clear, concrete visions of the team's purpose and success criteria so that everyone knows what they must do. By being persistent and enthusiastic, they also demonstrate a strong personal conviction to the team purpose. They show that conviction is based upon what they think is best for the project, not what they need to get their own "ticket punched" or to advance personally. Leaders generate commitment and enthusiasm by constantly "talking up" team purpose, objectives, and member roles. They recognize team members' technical competency and allow them leeway to use it. They delegate responsibility and allow members to take turns chairing meetings and directing the team. (Figures 13-27 and 13-28 illustrate

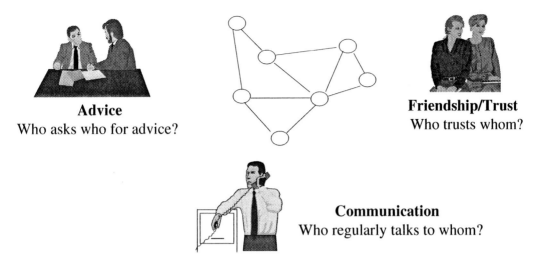

Advice
Who asks who for advice?

Friendship/Trust
Who trusts whom?

Communication
Who regularly talks to whom?

Figure 13-25: Three important qualities associated with IPT leadership

successful and unsuccessful delegation and sponsorship strategies.) They praise team achievements and are quick with compliments, thanks, or celebrations whenever something is completed or done well. When implementing the team based approach for the first time, key roles in the organization need to be politically mapped.

There are four key roles in managing the cultural change that will be required when improving engineering operations. And there are various ways to communicate the change. For example, the linear approach is to communicate the change top down. The staff approach is to communicate peer to peer. These are illustrated in Figure 13-26.

In the traditional linear top down approach, it is important to cascade the desired changes systematically down through the organization. This includes repeatedly targeting, coaching, and reinforcing. Unfortunately, when certain members of middle or executive management do not reinforce the change, a black hole forms. Information that reinforces the change is not passed on. The initiative fails, and this is illustrated in Figure 13-27. Figure 13-28 illustrates how to be successful when passing the change message on.

13.3.2 Charismatic, interpersonally competent, involved

The fact is, IPT leaders must convince others to get things done. They have to convince team members, customers, suppliers, and others. They need to be able to deal with people at a personal level and have the charisma and interpersonal skills to make people listen to them, respect them, and trust them. They must overcome the silos where information moves vertically through a protected organizational fiefdom and the hierarchy issues illustrated in Figure 13-29. When people trust

ROLE   RESPONSIBILITY

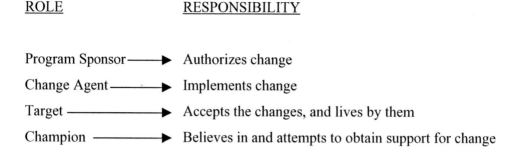

Program Sponsor ———▶ Authorizes change

Change Agent ———▶ Implements change

Target ———————▶ Accepts the changes, and lives by them

Champion ——————▶ Believes in and attempts to obtain support for change

THESE KEY ROLES ARE INTERRELATED

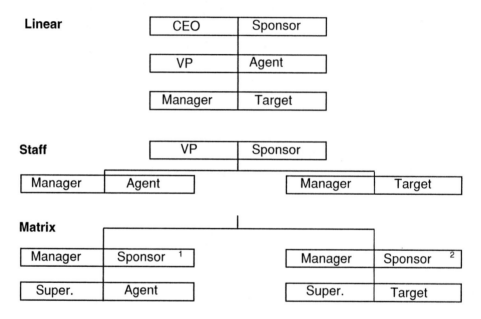

Figure 13-26: Key roles

and respect the leader, they are more likely to fulfill their duties, in part, out of obligation and a desire to not let the leader down.

The leader needs the technical competence to make informed decisions and to understand technical challenges, but technical competence is less important than being able to "lead" – to convey vision, gain commitment, and get people to pull together. Better than deep technical knowledge is broad knowledge and the ability to appreciate the range of viewpoints of all team members.

Project leaders sometimes avoid involvement in details lest it distract them from managing the big picture. In IPTs, however, not only is avoidance a near impossibility, but occasional involvement in details is desirable as a demonstration of the leader's interest in daily things that irk the team. Though quick to work next to the rest of the team, a good leader is also careful not to "meddle."

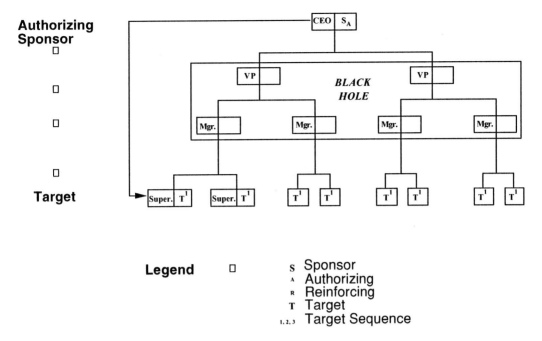

Figure 13-27: Unsuccessful sponsor strategy

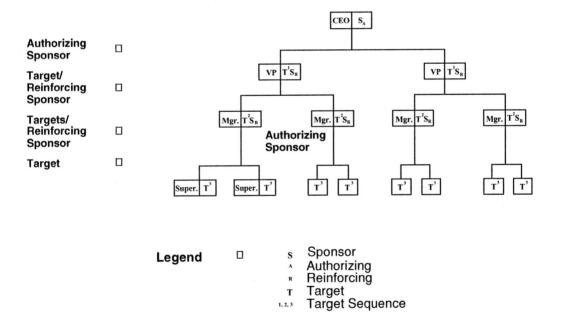

Figure 13-28: Successful sponsor strategy

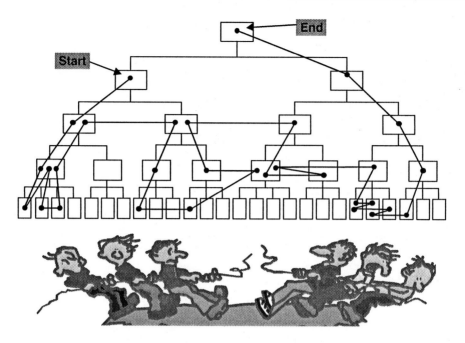

Figure 13-29: The silos and hierarchy are a big part of the barrier

13.3.3 Facilitate teamwork

IPT leaders are not the sole decision makers, but they facilitate groups of decision makers. Success on an IPT rests on the leader's being able to integrate effectively the expertise and opinions of all participants. The leader must know how to assist the group in developing behaviors that promote teamwork and to model the behavior through his or her own actions.

13.4 TEAM BEHAVIOR

Despite the pivotal role of the leader, there is much about teamwork that derives simply from the way the group itself behaves.

Obstacles to teamwork stem from, and are reinforced by, confusion about where the group is going, members' conflicting expectations, and dysfunctional group rules. These are issues that are best resolved very early in a team's existence, so that before work begins, members of the IPTs have an opportunity to discuss them and reach agreement. This is best done in the context of a two- or three-day meeting held off-site and facilitated by a trainer or someone skilled in facilitating group dynamics.

When people first join an IPT group, they bring expectations about what they will do and what others will do, among others. At this meeting, the leader and the members discuss their expectations. The team should take time to investigate in detail where there are disagreements and then work to reach a consensus on what the members should expect from one another, from the leader, and from the project. Meantime, members get better acquainted and begin to build trust.

Though members may already know about the project's purpose and goals, at issue is how they interpret them. After discussing each member's interpretation, the team arrives at a consensual definition of its purpose and goals. It then defines shorter term operational objectives and prepares a broad plan and work schedule.

The group will spend much of the meeting deciding how it will work as a team. All groups naturally develop tacit rules, called norms, which regulate group behavior. Some of these norms are detrimental to teamwork. To steer the group away from the detrimental norms and toward those that will benefit teamwork, the team develops a list of operating guidelines. Typical issues addressed by these guidelines are

- The decision-making process
- Communication and participation
- Responsibilities
- Meetings
- Conflict resolution
- Gripes

Some guidelines for questions include how decisions will be made and who will be involved. Which decisions will the team make? Which decisions will outsiders make? When will majority rule be used? When will consensus be used? How will the team encourage frank discussion? How will quiet people be heard? How will part-timers, outsiders, and other teams be kept involved and informed? How will responsibilities be assigned? What will be done when work falls behind? How will obligations beyond members' expertise be handled? How will the work load be distributed? When will meetings be held and how long should they take? Who will issue agendas and take minutes?

Conflict is also addressed. Conflict is inevitable, but the only productive way to handle it is through confrontation and joint problem solving. Guidelines might mandate, for example, openly discussing all disputes; confining disputes to the parties most affected; and initiating problem solving through mutual consent. By evaluating individual stances on an issue, a "stakeholder map" can be created. This is illustrated in Figures 13-30 and 13-31. In Figure 13-30, Z is highly influential and is blocking the need for change.

The guidelines should state how to encourage frank discussion. Though these will help keep gripes and complaints from growing out of control, they alone might not be enough. The team might want to assign a volunteer the role of team confidant, someone to whom people can take their gripes in confidentiality, and who then takes gripes to the leader or group for a resolution. Specific actions can then be taken to build support and achieve consensus.

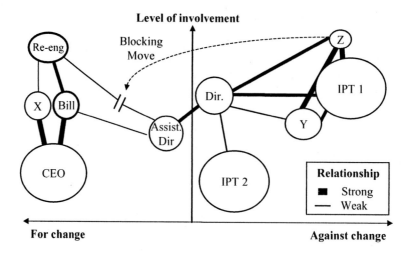

Figure 13-30: Map of support and resistance for an improvement initiative

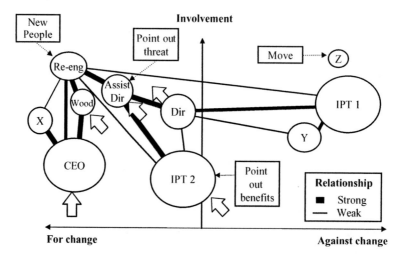

Figure 13-31: Removing the blocker

Figure 13-31 illustrates that the organizational relationships reconfigure when Z is removed. The organization exhibits elasticity, and the lines of influence change.

Not all the guidelines will be strictly followed; more often than not, the team will seldom refer to them again. However, simply having gone through the process of setting guidelines helps the team to avert conflicts, false starts, and bad work habits. Guidelines help get the team off to a better start.

No matter how well the project begins, there is almost always backsliding during a project. Someone in authority will resist the change or resist the team implementation goals. About once a week, the team should set aside time for self-reflection. For longer IPT efforts, a day-long session every several months might be needed. During this time, the group reviews its performance as a team and looks for emerging obstacles and prepares new plans and guidelines to overcome them.

13.5 CONCLUSIONS

Engineering improvement efforts are technological endeavors, but among the most difficult problems they face are organizational and interpersonal obstacles to effective teamwork. Paying attention to organizational politics, leadership, and group behavior considerations addressed here, most of these obstacles can be overcome. The considerable front-end effort this requires is worthwhile because it allows groups to begin sooner, and to continue thereafter, working as teams.

Implementing IPD – Lessons Learned Case Study

In this chapter, we will study an engine manufacturing company we will call ACME Engines as it begins to adopt a new integrated product development process based on concurrent engineering and integrated product teams around 1998. There are risks in implementing sweeping changes within an organization. One very real risk comes from thinking of the implementation in too narrow a context. One of the fundamental bodies of knowledge upon which this book is based is concurrent engineering (CE). Many managers will confuse a simple CE implementation with the holistic approach to engineering process management discussed in this book.

In implementing the changes, ACME observed that CE and IPT theories are being embraced across the aerospace and automotive industries. However, they also observed that implementation requires changes to how they organize, manage, and perform the engineering activities. These changes were not well understood nor accepted. As described in a *Fortune* magazine article, "The Trouble with Teams" (September 5, 1995), many other companies have also found the transition to teams to be much more difficult than expected. The director responsible for leading the changes at ACME emphasized, "Our biggest blunder was to underestimate the impact of organizational resistance to change." They indicated initially that the team approach may not seem to be as efficient as the old approach but that they are beginning to see now that it eventually provides superior results.

ACME observed common issues and lessons that need to be considered when implementing concurrent engineering practices and integrated product teams on the new engine program. They anticipate that other programs will struggle with these same issues during deployment. By recording these issues and lessons, we hope to assist and simplify your deployment and avoid some of the hurdles.

14.1 LEADERSHIP AND COMMITMENT

Let's review the background of ACME before examining the lessons they learned. ACME changed direction in 1995 with a major downsizing from 130,000

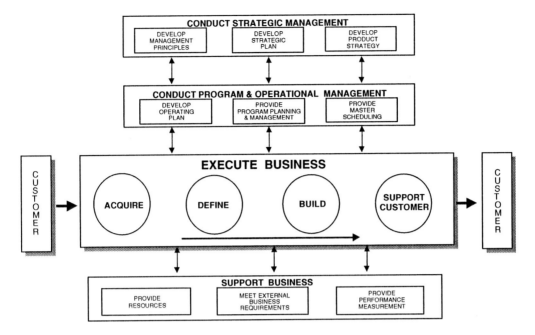

Figure 14-1: High level business process model

employees to 44,000 in 1996. They became market-driven, focusing on affordability and teamwork. Strategically, the company returned to focusing on diesel products. They reduced management layers and concentrated on shareholder value.

They undertook a major continuous improvement effort from 1996 to 2000 but found only multiple initiatives with no appreciable benefits. The performance efficiency improvement (PI) organization was formed late in 2000 to integrate the various initiatives and deploy the improvements to the program. PI is a deployment organization. Their initiatives included:

- Reengineering the product development process
- Enterprise resource planning
- Improve the "support customer" process
- Information technology infrastructure

They adopted a high-level business process model based on the value chain (see Figure 14-1).

The PI Program Director emphasized that the business process model must be under configuration control. As the ACME Engines change team began to apply the new methods developed in the continuous improvement program, it realized that its biggest challenge was getting people to adopt them. They had grossly underestimated human inertia. Most people paid lip service to the changes. The team even thought the human issues unworthy of their time.

At the beginning of ACME's transition, there was a strong commitment from the management of product engineering. Initially, this commitment was in the form of pilot projects, with support from external consultants. Later, it took the form of a policy statement from the director of a major new engine program that required the program to implement IPTs.

Even so, some people resisted the changes. To overcome this resistance, members of the IPTs needed to understand the concepts, see the benefits to the program, and understand the changes to their role. The IPT integration leader stated everyone supported the new approaches in theory as long as they did not affect them personally.

In many instances, leaders avoided or watered down the changes in response to pressure to deliver products, or because of uncertainty that the new methods would be as effective as the old methods, or from fear that the new methods would undermine management authority. For example, if a manager decides to review and approve all IPT decisions, the empowerment of the IPT is jeopardized. The team members and managers had a very difficult time giving up their functional hats.

Lesson 1 – Strong upper management commitment to drive the implementation of IPD and IPTs in the face of opposition is required. This must take the form of hands-on visible support, not just five-minute speeches.

When the new approaches were first adopted, everyone was optimistic but as they began to grasp all the changes required, confusion and doubt developed. People began to ask, "What is my role? Does the IPT have the authority to make this decision or do we need approval?" A crisis developed as the uncertainty grew, and the IPTs became focused upon the IPT schedule and products. Some people advocated going back to the old approach.

The confusion and doubt are a typical step in the progression of learning a new approach. At the beginning, they had no idea of an IPT structure. An IPT could include from 30 to 180 members. They found that the project leadership needs to continue to actively promote and personally use the new approach whenever possible. This may involve refusing to review and approve some decisions to emphasize the IPT empowerment or using consensus decision making to demonstrate commitment to the process. The project leadership needs to be the role model and realize that the IPTs will mirror the behavior of their leadership.

Lesson 2 – Three to six months after adopting CE and IPTs, there is a high level of frustration and a desire to revert to the familiar approaches. Strong leadership is required to continue. Strong leadership means getting involved hands on and demonstrating the new behaviors.

14.2 IPT SETUP

An IPT leader indicated that getting the IPTs off to the right start can prevent future problems.

The process begins with a clear product-oriented work breakdown structure for the project. The WBS needs to be closely related to the engine system being developed and the end items received by the customer. The WBS hierarchy needs to be organized along system, subsystem, and component lines. Setting up the WBS this way enables IPTs to be established to develop tangible deliverables. The WBS oriented to functional disciplines inhibit the use of IPTs.

A clear IPT purpose is required. The IPTs should be organized around end items of the WBS. The IPT mission and specific end items to be provided to the customer must be clearly specified. The cost, schedule, and technical requirements for the end items must be clearly defined. The relationship of the IPT's end items to other end items in the system must also be specified. This is often referred to as a deliverables architecture. This is an iterative process among the teams.

Identifying the customers of the IPT deliverables enables the teams to better understand the expectations for their products. Customer here refers to end users, external customers, and internal customers. It also helps to identify points of contact or liaisons needed to complete the team's work.

Each IPT must have an understanding of what constitutes success and how that success is measured. Business process and product measures need to be defined. The process and product measures should include cost, schedule, and technical measures. Process measures are such things as planned vs. actual completion dates for the specification outline and planned vs. actual costs. Product measures are such things as system failures, specification errors, and product cost.

Progress reports should be organized by the IPTs and their products, not by functional disciplines. The IPTs need to be able to measure and track their progress. When subcontractor or functional discipline reporting is needed, the subtotals should be shown by IPT and product to allow consistency with IPT reporting.

Incentives that encourage teaming need to be established. This may take the form of team incentives when IPT performance exceeds a defined level. Incentives should reward achievement of measurable goals.

IPTs should consist of cross-functional resources. The IPTs need to be composed of all the engineering disciplines necessary to develop the customer deliverables completely. IPTs that are missing one of the engineering disciplines are handicapped in trying to perform their mission. IPTs that include the customer and users are further enhanced. IPTs should assess their needed skill mix and attempt to obtain or grow the skills they lack. In addition, it is important to identify when resources are not needed. This avoids team members sitting around idle.

Functionally oriented teams, such as a test team, or a software team, eliminate many of the benefits of IPTs because the products are only part of a complete deliverable.

Team norms, a set of operating principles for the IPT that the IPT members agree to follow, need to be followed. They include such things as meetings, which will always have agendas and minutes. Decisions will be recorded and published, and team members will collaborate on and review work in progress.

The result of a properly set up IPT is that the team's scope of responsibility, empowerment, and resources are defined. Within this scope, the team has overall authority and need not seek approvals for its actions. This IPT scope allows product decisions and the work to proceed rapidly.

Lesson 3 – Take the time to define clearly the IPT purpose, end products, customers, process and product measures, resources, and team incentives. Encourage the IPTs to act within their empowerment scope.

14.3 DECISION MAKING

Consensus decision making allows all members to contribute to IPT decisions, but the consensus decision-making process at ACME encountered several problems. The people who make up the IPTs had little or no experience with consensus decision making. Most were classically trained mechanical and electrical engineers with little training in applied management skills. They typically had experience only with decision-making by majority or by individual. Many people did not understand that a consensus team decision may not have been optimal for each individual but must be workable. As a result, an individual may believe that it is their right to refuse to agree to a decision.

It is widely accepted that achieving consensus takes longer but pays off with better decisions. ACME found that until people learn how to use consensus decision-making methods, consensus decision-making takes much, much longer. They concluded that new skills and considerable practice are required to deploy consensus decision making.

The IPTs need to see consensus decision making work effectively. Unfortunately, on some teams, it was abandoned before the IPTs used it enough to become effective.

Second, people outside the IPT overruled IPT decisions. This was either because the team made decisions outside their responsibility scope or because others did not accept the team's mandate. Many times they found that the team was not properly set up; there was no clear understanding of the team's purpose and scope. When an IPT's decision was overruled, it had a paralyzing effect on the IPT, creating a cynicism that there is just lip service to the new approach. Both the team and the program leadership must understand and agree upon the team's decision-making boundary.

Third, IPTs tried to make every decision by consensus. Some decisions need to be made by an individual, some by small groups, and others by consensus. Decisions should be limited to the smallest group that they affect within the IPTs. ACME found that IPTs can usually agree on which decisions can be made quickly by the IPT leader and subteam and which require consensus of the IPT members.

A fourth problem developed when important decisions were made by edict or by a small group of individuals. This sends the message that consensus is not the real decision-making mechanism. For consensus decision making to succeed, it must be embraced and used by the leadership; otherwise, it will never be generally accepted by the teams and the program.

Lesson 4 – Carefully define the consensus decision-making procedure and when it is to be used. Use it to make some important decisions at all levels of the organization.

One approach to implementing this lesson is to raise the team's expectation that initially consensus decision making is difficult; using it takes discipline and practice. Try to involve everyone in defining and improving the decision-making procedures.

Two examples of basic decision-making procedures being used by ACME are "thumb voting" and multivoting. Thumb voting allows everyone three alternatives: if you favor the decision, put your thumb up; if you can live with it, put your thumb to the side; and if you cannot live with it, put your thumb down. When there is a thumb down, the team takes time to understand the concerns and synthesize a new alternative that everyone favors or can live with. Calling for a visible thumb vote ensures an understanding of everyone's viewpoint and builds commitment to the decision.

Multivoting is used to select and prioritize from a brainstorm list. Usually each member of the team is given a number of votes to identify the items that are most important. The group selects the items with the most votes. All items from the list are retained. Multivoting helped the team reach agreement on actions and directions.

ACME found that when these simple decision-making procedures were practiced, they built consensus and effective multidiscipline views.

14.4 ROLES AND RESPONSIBILITIES

ACME learned that people's roles and responsibilities change with the implementation of concurrent engineering and IPTs. This change is difficult because the old roles and responsibilities are deeply ingrained. In the new environment, roles and responsibilities are often very different. Unless the new roles and responsibilities are defined and documented, each member of the IPT may have a different understanding of the roles. Old responsibilities with titles such as team lead or manager contribute to this confusion.

For example, ACME considered the role of the team lead. In the past, many decisions were left to the manager. A lead came up with the ideas and recommendations, but the manager had the final say. The lead expected the manager to bless a decision before acting upon it. In the new environment, the IPT has the final say, but many members of the IPT may still expect a functional manager to bless decisions.

Managers who still wanted to approve ideas before the IPTs committed to implementation posed another pitfall. The result was confusion, with some people still operating under the old system and some under the new system. The roles of the managers, IPT leads, and customers are particularly difficult to define and understand.

When the roles and responsibilities are confused, the risk of abandoning the new approach is greatest, and some justify doing so on the basis that it does not work as well as the theory. This is a critical point. ACME's efforts were excellent in the development stage. The theory was well defined and documented. However, the implementation effort was totally underestimated because it was led by people trained only in an engineering discipline with little understanding of organizational change management.

> Lesson 5 – Make sure the leadership (including all levels of management) and the IPTs define, record, and commit to the new roles and responsibilities. Periodically, the leadership and the IPTs should review and revise the roles and responsibilities.

The discussions that followed on roles and responsibilities allowed everyone to understand the different expectations for a particular role. People who still want to operate under the old system are identified. New expectations for each role are established. Target coaching and influencing occurred, but they occurred after deployment.

14.5 COMMUNICATION

Communication is more than brochures, posters, and get-togethers. In the team environment, effective communication is very important. A common misconception that the IPTs encountered was the expectation that most work is now performed together. Questions and decisions that in the past were addressed by a single person were now discussed by the IPTs. The result was long periods of discussion in which little was accomplished. ACME indicated that the teams found this particularly difficult.

ACME learned that not all work is done together by the team under the new process; some work is done best by subteams and individuals. The IPTs need to be careful to decide which work is best done by the entire IPT, by a subteam, or by a team member.

When IPTs or subteams are doing work, everyone has to be involved in meeting the objectives, controlling diversions, and seeking to understand one another's views. Team members must exercise a great deal of personal discipline to speak only when they have something to add and to relate what they say clearly to the point of discussion.

One of their teams developed a does-it-matter team norm to remind themselves to focus their communication. Another common team norm is to limit war stories to one minute and debates to five minutes. After these limits are exceeded, the discussion is taken off line.

> Lesson 6 – Effective and efficient team communication depends upon the IPT membership recognizing which work is best done as a team, as a subteam, and as individuals.

After a short period, IPTs began to practice internal communication. However, communication difficulties between IPTs did arise. The IPTs became focused on their own objectives and products, to the exclusion of the other IPTs. The overall engine program suffered when inter-IPT communication was not addressed.

Ideally, one may think this is the role of the integration team. However, practically speaking, it is not possible or desirable for the integration team to address all the interteam issues. It is better to have the IPTs resolve for themselves as many problems as possible.

One of ACME's teams created a liaison role to address inter-IPT communication. A liaison was chosen for each of the other IPTs to identify dependencies, track progress on the IPT dependencies, and lead resolution of inter-IPT issues.

Another project created a team of IPT leads to handle inter-IPT issues, communication, and plan adjustments. This team was able to resolve many IPT issues and actions and thereby reduce the number of issues that had to escalate up the organization for resolution.

> Lesson 7 – Establish a formal mechanism for communication between IPTs, and identify IPT dependencies early.

14.6 TEAM SKILLS AND TRAINING

With the formation of IPTs and process integration framework, ACME encountered four major skill-related difficulties. The first difficulty was a lack of understanding of other engineering disciplines. For the IPTs to be effective, all team members needed to understand a little about the life-cycle steps of other disciplines. Cross-discipline understanding allowed IPT members to contribute to each other's work. ACME has a very structured process framework that clearly defines the life-cycle phases, maturity gates, and detailed processes. Despite the structured process methodology, teams still encountered difficulty because of ignorance of other disciplines. Previously, the functional organization worked against the development of broader cross-discipline skills.

A second difficulty came from a lack of interpersonal team skills. Effective communication, listening, encouragement of other members, and suppression of individual egos are examples of some of the skills needed to be an effective member or leader of an IPT. The lack of these skills led to ineffective decision making, ineffective meetings, and inferior IPTs. ACME found that after a while, some IPTs began to identify their training needs.

A third difficulty was not using team-building methods (consensus decision making, facilitation, brainstorming, etc.) enough to employ them effectively. Training should allow the IPTs to practice team-building methods on real problems. Most people have to see and practice these methods many times before truly adopting them. The program director emphasized that a message has to be stated over and over to stick.

Finally, the broader responsibilities assumed by all members of the IPTs require basic project management skills. Team members require cost and schedule estimating and tracking skills.

ACME has increased off-the-job training from 80 hours to 170 hours. Most of the training concentrates on interpersonal and management skills.

Lesson 8 – Make sure that the IPTs are supported with training that defines a core set of engineering discipline skills, interpersonal skills, team member skills, and project management skills.

A changing work sequence to develop engineering products

In the past, ACME engineers have developed their engineering deliverables through a process of informal consultation and individual creation. After the initial creation, the engineering deliverable is subjected to a series of reviews and reworks until the customer accepts it. The views of different engineering disciplines were often reworked into the deliverable through repeated reviews. This cycle of review and rework was both costly and time consuming. Invariably, the views of the different engineering disciplines were only partially incorporated.

In a concurrent engineering environment, ACME found that the sequence of engineering product development changes. All disciplines necessary to create the product are identified and assigned during creation of the deliverable. The development process now involves creating subteams with a mandate to review and comment continuously.

Subteams use brainstorming, consensus decision making, and other creativity techniques. For example, brainstorming was used to list the design characteristics from all the relevant disciplines as guidance to the individuals creating portions of the work products. The subteams and individuals iteratively refined and agreed to the deliverable content as it was being created. The result is collaboration with shared ownership of the engineering deliverable. In essence, the work sequence is reversed with the review occurring largely prior to the creation.

Implementing concurrent processes effectively requires an environment that enables everyone to have immediate access to engineering deliverables as they are being created. Even with this environment, ACME were most successful when they colocated the team including subcontractors, customers, and users and provided everyone access to all the deliverables under development.

ACME encountered several problems when introducing concurrent processes. First, some subteam sessions that were not focused on developing a specific deliverable wasted time. They found that the most effective subteam sessions needed to define specific objectives and to use brainstorming and consensus decision-making procedures. These sessions were followed by a consolidation step to publish the session results. The subteams also had to train themselves to not constantly revisit decisions made in previous sessions.

Second, ACME engineers were surprised and overwhelmed with the number of comments received during the creation step. Many of the comments were extremely useful, but many others were not. The engineers now expect more comments early in the creation step and are prepared to deal with them. Additionally, they have disciplined themselves to think through comments and only pass on those that matter.

Third, they did not recognize that some of the review and rework sequences should be eliminated from the start. They found that the IPTs and their customers must be given the expectation that the creation sessions at the beginning will reduce the reviews and rework at the end. Otherwise, the result is a longer creation step with a review and rework step that takes the same time or longer.

> Lesson 9 – Engineers and managers need to recognize and adopt a different workflow approach to develop engineering deliverables to realize the benefits of concurrent engineering.

A balanced systems approach to IPD and IPTs

ACME learned that the implementation of the new process interacts with the overall system of management and personnel practices. IPTs often challenge the empowerment boundaries and the appraisal and reward systems.

An IPT will question the scope of its empowerment. It is normal for the initial scope to be defined incompletely. Sometimes an IPT will make a decision outside of its scope, or someone not on the team may make a decision within the IPT scope. In either case, care must be taken to explain the basis for changing the decision to make sure that an IPT approach is not undermined.

Following are some common ACME IPT questions about the appraisal and reward systems:

○ How is the individual appraisal related to the success of the team? Should not everyone on the team get the same appraisal?
○ Don't individual appraisals and rewards encourage each member of the IPT to try to get individual credit for team activities?
○ Why doesn't the IPT or the IPT lead do the appraisals? They are most aware of each team member's contributions.

Lesson 10 – The new approaches require integration into the overall system of management, with a focus on establishing the IPT empowerment and determining how performance appraisals and rewards will be administered in the team environment.

14.7 CONCLUSION

Implementing IPD and IPTs requires significant changes to the way we organize, manage, and perform work. The management needs to offer a strong commitment to ensure the transition. The new skills require practice and time for the teams to use them effectively.

The key lesson is that developing the processes and software and documenting the new systems approach are relatively easy compared to deploying it to an existing organization. The existing organization has its own behaviors, rules, work methods, and power structure and ignoring it will result in a failed effort. This happened at ACME and resulted in a "recovery plan." Recovery plans always cost two to five times as much as deploying right the first time.

Initially, there was a great deal of optimism, but as reality hit, there was danger of creating frustration and cynicism. To avoid this, IPTs need periodically to assess their progress and plan improvements. The process needed an outside facilitator to assist and advise the IPTs.

Part 4

Appendixes

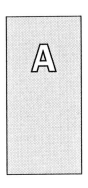

IPD Maturity Self-Evaluation Tools

A.1 PURPOSE OF THE EVALUATION TOOL

This section contains an IPD Maturity Self-Evaluation Tool for assessing the status of integrated product development best practices on a program, as well as the overall maturity in IPD within a business. The results an be used to provide feedback to program managers, integration team members, integrated product team members, functional managers, and other site IPD stakeholders for process improvement activities. The survey can also help to identify new best practices in IPD by sharing lessons learned.

A.2 SCOPE OF THE EVALUATION TOOL

The evaluation tool/survey appears in Table A-1. It is structured in the form of statements of reality rather than questions of perception. It is also important to note that duplication between the sources has been eliminated, resulting in a single set of statements. Therefore, to receive an accurate assessment, a response to all the statements, regardless of their origin, is necessary. Additionally, at the end of the statements the assessor is given the opportunity to express subjective opinions, comments, and suggestions about the implementation and effectiveness of IPD in an essay-type format.

The user may choose to use this tool as a *checklist* to determine what IPD techniques are or need to be implemented; as a *survey*, responding to the statements in a "yes/no" fashion; or as a *maturity indicator*, where a specific response to each statement contributes to an overall measure of maturity.

When the tool is used as a maturity indicator, to the right of each statement are five possible responses to the subject/description/action addressed in the statement:

Almost Always (AA)
Some Times (ST)
Almost Never (AN)

Don't Know (DK)

Not Applicable (NA)

(See Table A-1 for response interpretations/maturity scale projections.)

○ Check the Almost Always (AA) box if the statement describes a characteristic or an action that very frequently exists or is in place on the IPD project to which you are assigned.

○ Check the Some Times (ST) box if the statement describes a characteristic or an action that is not consistently apparent or happening on the IPD project to which you are assigned.

○ Check the Almost Never (AN) box if the statement describes a characteristic or an action that very seldom exists on the IPD project to which you are assigned.

○ Check the Don't Know (DK) box if you are not sure whether the statement is even applicable to the IPD project to which you are assigned.

○ Check the Not Applicable (NA) box if the statement is not applicable to the IPD project to which you are assigned.

A.3 ADMINISTERING THE MATURITY SELF-EVALUATION SURVEY

The maturity evaluation survey should be administered at several points during a program, beginning with the proposal phase and ending with product delivery. This is intended to give an indication of program/business IPD maturity trends and identify specific areas for improvement. Typically, the program team members, including the program manager, IT members, and IPT members, are interviewed. The tool is most effective when administered by an individual from outside of the program, to ensure impartiality and anonymity in the interviews. This individual may be the IPD facilitator/champion for the site or some other appropriately designated representative. When used as a maturity indicator, a composite of the results from the evaluation should be provided to the program team for feedback and/or corrective action planning.

The tool has a very broad scope of application and as a result can be used at a company, program, and/or IPD team level. To ensure that a meaningful measurement is achieved, recommended evaluation sample sizes are provided in Table A-2. These recommended survey sample sizes are based on the survey group population and the desired evaluation precision.

The responses to the evaluation survey subject/description/action statements will indicate the scale of maturity within each category or level of maturity in IPD. Table A-3 provides guidelines for correlating the survey responses to a scale of maturity within each level of maturity. For example, an excellent rating within

Table A-1: IPD maturity self-evaluation survey

Maturity level	Key tenet	Statement	Response				
			AA	ST	AN	DK	NA
Level 1	Management commitment	IPD philosophy is adopted and practiced by management, staff and personnel					
		IPD philosophy is visible internally to employees and externally to customers and suppliers					
Shared philosophy	Empowerment	IPD policies and procedures are in place and followed by program(s)					
		Business processes are adapted to IPD methodologies					
		Roles and responsibilities are defined					
(IPD approach adopted)	Customer focus	Team members have received IPD training					
		Team leaders have received team leadership training					
		Teams are given the authority and resources to manage the product					
		Team members understand their roles and responsibilities					
		Teams accept the responsibility and accountability for their results					
		Team plans have been developed and are in use					
		Decisions are made at the lowest level commensurate with the risk					
		IPD philosophy is adopted and practiced by customer					
		IPD philosophy is adopted and practiced by suppliers					
		Customer is involved in all aspects of product and process development					
		Awareness of customer's needs					
Level 2	Communication	Standard practice of the integrated product team is to communicate with the following:					
		Customer					
		Engineering					
Good information transfer		Manufacturing					
		Sourcing					
		Suppliers					
		Test					
		Quality/product assurance					
		Logistics					
(people working together)		Sharing of data and information within and between IPTs is standard practice					
		Colocation is established for core IPT members					
		Support members are provided facilities within the program area					
		Virtual colocation is established for off-site IPT members (customer, suppliers, etc.)					
		Team meetings are held on a basis consistent with the needs of the program					

(Continued)

Table A-1: (Continued)

Maturity level	Key tenet	Statement	Response				
			AA	ST	AN	DK	NA
		Team meetings are attended by the entire team					
		Informal meetings to address specific issues are held on an as-needed basis					
		Communication tools are in place to provide interchange of data and information					
		Program has a standard information distribution process					
		IPTs coordinate with other teams, when necessary					
		Team leaders are involved in technical and program issue resolution					
		Technical and program status is communicated between IPTs and functional areas					
		IPTs and support members have the technical skills required to perform their tasks					
		IPTs and support members have the non-technical skills required to perform their tasks					
		Available lessons learned are accessed					
		Lessons learned are shared with team members and functional areas					
		IPT performance is measured against standards					
		IPT performance is recognized and rewarded					
		Customer has agreed to the IPT performance measurement standards					
		Team Leaders participate in the performance reviews of IPT members					
Level 3	Multidisciplinary teamwork	All necessary functions are represented on the IPT					
Multifunctioning teaming		All necessary functions participate in the IPT decision-making process					
		Teams have active customer representation					
(teams organized and functioning)		Teams have active subcontractor representation					
		Teams have active supplier representation					
		IPT skill mix is supported by functional managers					
		IPT members understand and follow team plans					
		IPT plans have program and functional management approval					
		IPT plans are reviewed regularly for progress toward commitments					
		Team decisions are based on input of the entire IPT					
		Team members make decisions and act in the best interest of the IPT					
		Key decisions are brought to the IPT and approved by consensus of the team members					
		IPT decisions are supported by program and functional managers					
		Functional areas are responsive to requests for resources and information					
		IPT members have access to cost and schedule data					
		Manpower planning is driven by an integrated master plan					
		Funding profiles support multifunctional manpower planning					
		The program organization chart reflects an IPD approach					

Level 4	
Tools and processes operational (tools and techniques)	
Early and continuous life	An IPD implementation process checklist, or equivalent, is used
	Proposal assumptions and risks were used in the program planning phase
	Multifunctional basis of estimates (BOE) were developed in accordance with IPD handbook
Event-driven scheduling	A work breakdown structure (WBS) was developed in accordance with IPD handbook
	A single number tracking system is used
	IPTs have developed integrated statements of work in accordance with IPD handbook
	Key program events are defined
Seamless management tools	An integrated master plan (IMP) was developed in accordance with IPD handbook
	Customer and supplier inputs are included in the planning
	Customer scope changes are processed in a team environment
	A systems engineering management plan (SEMP) was developed in accordance with IPD handbook
Concurrent development of product and process	A manufacturing program plan (MMP) was developed in accordance with IPD handbook
	A sourcing program plan (SPP) was developed in accordance with IPD handbook
	A quality program plan (QPP) was developed in accordance with IPD handbook
	Plan changes are agreed upon and communicated in team environment
	Applicable aspects of the product life cycle are considered during development
Ecourage robust design and improved process capability	Plans are reviewed and revised when necessary throughout the program life cycle
	Key program events are time-phased
	An integrated master schedule (IMS) was developed in accordance with IPD handbook
	Every event in the IMP is included in the IMS
	Program critical path activities are identified and communicated to team(s)
Proactive identification and management of risk	Interdependencies with other programs are identified in the IMP IMS, and risk assessments
	IPT milestones are defined, planned, and scheduled by the team
	IPT agreements, or equivalents, have been established and agreed to by team members
	Assumptions, risks, and BOEs identified in the proposal phase are used during the development phase
Maximize flexibility for optimization and use of contractor unique approaches	Program cost, schedule, and technical parameters are defined, communicated, and tracked
	Product cost, schedule, and technical parameters are defined, communicated, and tracked
	Manufacturing and support processes are developed concurrently with the product
	Processes are benchmarked as an essential practice of the functional areas in support of IPTs
	EPI discipline processes are used by the IPTs
	Functional processes (other than EPI processes) are used by the IPTs
	CAE tools are used by the IPT(s) to optimize product performance and producibility
	Design reviews and trade studies are used to optimize product design

(Continued)

Table A-1: (Continued)

Maturity level	Key tenet	Statement	Response				
			AA	ST	AN	DK	NA
		Design reviews include customer representatives as well as functional personnel, as req'd					
		Design-to-cost (DTC) tools, techniques, and methodologies are utilized by the IPTs					
		Producibility tools, techniques, and methodologies are utilized by the IPTs					
		Manufacturing process control techniques are utilized					
		Product and process prototyping and/or validation is utilized by the IPTs					
		Information is accessible for IPT decision making					
		Tools for multi-functioning access and exchange of information are available					
		Risk Identification is performed by teams					
		Risk mitigation plans have been developed and implemented					
		Product and process maturity are incrementally demonstrated					
		Acquisition reform initiatives have been implemented, where applicable					
		Openness to supplier solutions and commercial best practices is implemented, where practical					
		Regular program status reviews are held for all team members					
		Applicable program and product metrics are utilized					
Level 5	Continuous improvement	Customer needs drive program planning and execution, and product performance					
IPD institutionalized		Formalized and documented IPD methodologies are utilized on all programs					
		IPD methodologies are imbedded in the functional processes					
(mature implementation of IPD)		Key suppliers utilize IPD methodologies within their companies					
		Product and supporting processes are optimized to meet customer needs					
		IPD metrics are identified and monitored					
		IPD metrics are compared to benchmarks to measure improvement					
		Status of metrics is presented at program reviews					
		Metrics data is fed back to functional areas for continuous improvement					
		IPD policies, procedures, and methodologies are continuously improved to reflect LL and BP					
		IPD evaluation survey opinions, comments, suggestions, observations					

1. How can IPD be improved on *your* program?
2. How can IPD be improved on *all* your programs?
3. What good things have you observed on your program?
4. What barriers have you encountered that prevent continuous improvements in the implementation of IPD?
5. Share any other comments, suggestions, or observations.

Table A-2: Sample sizes to achieve various tolerances at a 90 percent
confidence level

Population size	Precision (tolerance)					
	5	6	7	8	9	10
50	43	40	37	34	32	29
100	74	66	58	52	46	41
150	97	84	72	63	54	47
200	115	97	82	70	59	51
250	130	108	89	75	59	54
300	143	116	95	79	63	56
350	153	123	99	82	66	57
400	162	128	103	84	68	58
450	169	133	106	86	70	59
500	176	137	109	88	71	60
550	182	141	111	89	72	61
600	187	144	113	90	73	61
650	192	146	114	91	74	62
700	196	149	116	92	74	62
750	199	151	117	93	76	63
800	203	153	118	94	76	63
850	206	154	119	94	77	63
900	209	156	120	95	77	63
950	211	157	121	96	77	64

Table A-3: IPD maturity scale

Scale	IPD evaluation tool response percentages		
	Almost always	Some times	Almost never
Excellent	95–100	<5	—
Satisfactory	85–95	—	<10
Marginal	70–85	—	<25
Unsatisfactory	<70	—	>50

the maturity level 1 (shared philosophy) category may be accompanied with only a satisfactory rating in the maturity levels 2 and 3 (multifunctional teaming) categories. This type of finding may be indicative of a site or program in the early stages of evolution in using IPD methodologies. Ideally, however, the goal is to achieve an excellent rating in each of the five categories/levels of maturity in IPD. This scaling system not only highlights the degree of success achieved within each of the five categories of maturity but also provides visibility into and focuses attention on the areas that should be targeted for improvement.

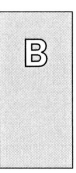

Chapter 2 Tables

Table 2-1: IPD phase objectives and deliverables example

Phase	Objectives	Deliverables
D1 Develop conceptual definition	The develop conceptual definition phase is required so that adequate technical objectives are established for the program. The conceptual design phase terminates at the successful completion of the specification review. The specification reviewer asks the question, "Is there sufficient technical definition to develop a program plan?"	O Conceptual definition package
D2 Prepare for management review and approval	The conduct management review and approval phase creates the program plan. It defines schedule, cost, scope, statement of work, and resources to achieve the technical objective and to access the capability of completing it. This process terminates in a launch review at which the reviewer answers the question, "Does this program show adequate return on investment?" or "Does the customer accept the quality, price/cost, and schedule?" The launch decision is equivalent to the final management approval meeting. The management proposal must address the five principal structures of the integrated enterprise.	O Launch review package
D3 Develop preliminary definition	The develop preliminary definition phase refers to the initial concepts defined in D1 and develops them so that the product definition is cohesive and complete. Design for manufacture, design for assembly, and life cycle cost analysis techniques are used in this phase to optimize the details of the design. This phase terminates in a preliminary design review, at which the design is shown to be sound, the financial model matches the design, and there is a high degree of confidence in the design as defined. The configuration is now frozen.	O Preliminary definition package

Table 2-1: (Continued)

Phase	Objectives	Deliverables
D4 Develop detailed product definition	The develop detail definition phase defines the configuration frozen in D3 and develops it in detail to the point where drawings can be produced. This phase terminates in a critical design review, at which time the designs are reviewed in detail and the acceptability of the design is confirmed. After the critical design review, the detail design is frozen.	○ Critical design review package ○ Design freeze confirmation
D5 Release product definition	The release product definition phase formally produces and releases the detail design developed in D4. This process involves two stages: (1) formal communication to the shop for part production (via data set) and (2) production of the technical definition of the product regulatory records for inspection purposes. This process terminates at the release of the product data models in compliance with the engineering practices and procedures.	○ Product definition package ○ Product data set ○ Technical definition for regulatory bodies
D6 Certify product	The certify product phase produces the documentation necessary to show that the product complies with the airworthiness or other standards and is in accordance with the original design intent. The certification phase terminates with the qualification of the design and the ability to enter the product into the marketplace.	○ Regulatory approval ○ Product qualification ○ Product documentation ○ Compliance program
D7 Complete program	The complete program phase includes performing a postprogram development review. In conducting this review, management seeks to identify and document best practices as part of a continuous improvement initiative.	○ Program completion review package
D0 Manage program	The program management architecture includes all of the planning, organization, management, reporting and corrective action activities that are product specific. These include general project management, resource planning and management, financial reporting, schedule reporting, and project specific performance measurements. It includes a program management methodology necessary to initiate D1 through D7 processes and to monitor and conclude the major milestone reviews of D1 through D7. The five subprocesses of D0 are repeated in each of the seven major phases of the IPD process architecture.	○ Program plan

Table 2-2: IPD phase descriptions and deliverables example

Phases	Subphases	Deliverables
D1 Develop conceptual definition	D1.1 Establish customer needs In this subphase, the marketing group identifies through market research what existing and future customers desire. Market research can take the form of one-to-one interviews, mass study surveys, etc.	○ Requirements ○ Definition report
	D1.2 Establish competitive products Establishing competitive products involves working with marketing to establish realistic product requirements (market requirements and objectives) with respect to the competition and available technology. It requires conceptual design studies and benchmarks to define competitive products and evaluate technology trends and development requirements.	○ Advanced design study report ○ Marketing requirements and objectives report ○ Benchmark report

	D1.3 Create strategic technology base
	Creating a strategic technology base involves collecting and defining the strategic technology available for incorporation in the product. It also identifies and controls work on additional strategic technology developments needed to improve the technology base available to meet market requirements. It establishes the manufacturing technology available and required to meet market requirements with respect to cost and weight. The prime aim is to develop and incorporate technology as it improves the product's functional capability, safety, and competitiveness or offers an advantage to the company without undue risk.
		○ Strategic technology report
		  ⮑ Materials/process
		  ⮑ Manufacturing strategy
		  ⮑ Computational methods
		○ Five-year strategic technology plan

	D1.4 Create product functional requirements Creating the product functional requirements process takes all the requirements and objectives from marketing, the regulatory authorities, and the company, augmented by the strategic technology base and translates them into product-related requirements and objectives. They do not define the physical product but instead define the product's required capabilities and features.	○ Product/process design requirements and objectives ○ Design standards manual ○ Drafting manual
	D1.5 Create product specification Creating the product specification process defines what the product is. It contains all the physical attributes of the product and the subspecifications for the subcomponents of the product. It also includes a preliminary optimization of the conceptual design.	○ Product specification document ○ Production requirements capability plan ○ Regulatory compliance plan ○ Technical requirements document

Table 2-2: (Continued)

Phases	Subphases	Deliverables
D2 Prepare for management review and approval	D2.1 Develop project plan	○ Project schedule ○ Human resource plan ○ Financial plan ○ Manufacturing plan ○ Program partnership plan
	D2.2 Prepare project infrastructure requirements Preparing the project infrastructure requirements involves analyzing the project plan and regulatory requirements to create the business philosophy, information structure, computing technology structure, physical structure, and business processes requirements that the project will require to meet the deliverables of the project.	○ Business philosophy ○ Product information structure ○ Work breakdown structure ○ Preliminary product structure ○ Database structure ○ Computing requirements plan ○ Project software standards ○ Facilities plan ○ Program-specific business process architecture
	D2.3 Plan infrastructure changes Infrastructure deals with the internal operations of product development. The elements include people processes, technology, information and organization. Each of these enterprise elements must be aligned and be in balance with one another. For example, it is inefficient to implement new CAD technology without aligning the business processes with the new information flow.	○ Infrastructure plan ○ Configuration management plan
D3 Develop preliminary definition	D3.1 Ensure product functional compliance Ensuring product functional compliance ensures that the product as defined in the model specification complies with the functional requirements. Technical design and analysis work and testing are performed to meet the configuration freeze objective. D3.2 Optimize product definition Optimizing product definition takes the product as defined and applies design for manufacture, design for assembly, and life cycle cost analysis techniques to optimize the product design. The design is also refined to optimize the performance attributes of the product. Further overall optimization and integration of the preliminary design is accomplished.	○ Systems architecture ○ Systems layout ○ Systems performance spec ○ Product geometry ○ Structural interfaces ○ Internal structural loads ○ Matrix of preliminary definition vs. product spec ○ Test report ○ DFMA worksheets

(Continued)

Table 2-2: (Continued)

Phases	Subphases	Deliverables
	D3.3 Confirm product configuration Confirming product configuration involves drawing conclusions from the various studies conducted in D3.2, packaging the outcomes and marketing the results to top management. Agreement is then achieved to implement changes.	○ Product measures ○ Trade study documentation ○ Preliminary design report ○ Updated preliminary product definition ○ DFMA reports
	D3.4 Confirm product financial assumptions Confirming product financial assumptions takes the current configuration of the product and assesses the estimated completion cost established in D2 to complete.	○ Updated program plan report ○ Test reports
D4 Develop detailed product definition	D4.1 Create detail product data Creating detail product data generates the detail product data which will eventually be input to the manufacturing system. These data are the basis for the analytical and test verification of the detail design and the basis for the initial certification meetings with regulatory agencies.	○ Part and assembly definition ○ Manufacturing data specs ○ Vendor data specs ○ Configuration audit ○ Integration reports
	D4.2 Confirm detail configuration by analysis Confirming detail configuration by analysis takes the detail configuration and demonstrates compliance to the functional requirements and the model spec by analytical means in enough detail to allow the design to be frozen.	○ Matrix of detail definition vs. product specification
	D4.3 Confirm detail design by test Confirming detail design by test takes the detail configuration and demonstrates compliance to the functional requirements and the engineering model spec by development tests in enough detail to allow the design to be frozen.	○ Test reports
	D4.4 Define means of regulatory compliance Defining means of regulatory compliance involves initiating certification activities that take the detail product definition and organize it in a form suitable for presentation to regulatory agencies. Initial type board meetings are held to confirm that the design may be frozen.	○ Draft compliance plan ○ Analysis of preliminary definition ○ Design approval review package

Table 2-2: (Continued)

Phases	Subphases	Deliverables
D5 Release product definition	D5.1 Obtain design approval Obtaining design approval involves collecting and ensuring the completeness of the detail models and drawings, data from preliminary design, tests, manufacturing, and logistics. In addition, this activity guarantees that the detail design and entire data have been reviewed and approved by authorized personnel of all disciplines and that the traceability of design decisions to the design and performance requirements is confirmed and completely documented.	○ Signature of design approval board ○ Preliminary release package
	D5.2 Release and control drawings and data models Releasing and controlling drawings and data involves formally processing the many forms of data through the release system, which contains the checks and balances necessary to communicate new information to the users, void existing or obsolete information, and control legitimate data.	○ Bill of materials ○ Process sheets ○ Assembly manuals ○ Tool design drawings ○ Solid and wireframe models ○ Two-dimensional drawings
	D5.3 Report design release status Reporting design release status utilizes the final design drawings, data, product design, documentation, parts list, geometry, spares, ground support equipment design, manufacturing information, actual release dates, tool coordination data, and material requirements from D5.2. This final activity of the detail design effort provides current status of a released product baseline definition; this also includes providing status information into the project management process.	○ Drawing at revision status ○ Physical configuration audit report
D6 Certify product	D6.1 Verify product functionality Verifying product functionality takes the finished product and performs sufficient testing to ensure that the market requirements have been met.	○ Component test report ○ Product test report ○ Systems test report

(Continued)

Table 2-2: (Continued)

Phases	Subphases	Deliverables
	D6.2 Perform product certification testing Performing product certification testing involves performing the required certification testing to comply with the regulatory requirements when different than functional testing.	○ Certification test report ○ Vendor qualification reports
	D6.3 Produce proof of compliance Producing proof of compliance produces the data and documentation that substantiates the statement of compliance.	○ Approved mod sheets ○ Certification report ○ Statement of compliance
	D6.4 Produce product documentation Producing product documentation produces all the documentation that the end customer/operator sees and uses to operate and maintain the product.	○ Operating manual ○ Quick reference manual ○ Service bulletins ○ Maintenance and repair manual
D7 Complete program	D7.1 Confirm market acceptance The marketing department will perform a postprogram review of the customer's acceptance of the product. Strengths, weaknesses, and opportunities for improvement are documented.	○ Marketing compliance matrix
	D7.2 Conduct closing financial review Senior management assisted by the finance department perform a detailed analysis of the various financial ratios on the program. The key elements include profitability, return on investment, return on equity, and overhead ratios.	○ Financial performance report
	D7.3 Document lessons learned This is performed for various program elements. Elements include business processes used in product development, manufacturing, and support. Other elements include organization structure, team dynamics, and technologies applied.	○ Lessons learned report for various elements
D0 Manage Program	D0.1 Initiate phase Initiating the phase establishes the means for authorizing the program.	○ Budget structure ○ Parametric estimates and forecasts ○ Technical requirements document
	D0.2 Plan phase Establishing the planning phase is focused around the development of structure for the program. It requires that the setup issues are addressed primarily in support of D1 and D2 deliverables.	○ Program management ○ Statement of requirements

Table 2-2: (Continued)

Phases	Subphases	Deliverables
	D0.3 Execute phase The execution phase establishes the means for monitoring, reporting, and taking corrective action for the project, throughout the IPD process architecture and also throughout the build and support processes. This is the primary process for carrying out the project plan. Within this process, the program team must manage the various technical and organizational interfaces that exist in the project.	○ Actual hours vs. budget hours ○ Actual dollars vs. budget dollars ○ Schedule compliance
	D0.4 Review and control phase The review and control phase monitors the project plan in relation to the cost, quality, and schedule previously determined in the project initiation. It ensures that the objectives are met by measuring progress and taking corrective action when necessary, throughout the IPD process architecture and also throughout the build and support processes.	○ Business philosophy ○ Quality visibility charts ○ Engineering change request work statement ○ Change request documents
	D0.5 Wrap-up phase The wrap-up phase formalizes acceptance of the project and brings it to an orderly end.	○ Closed contract

Table 2-3: Customer deliverables/definition example – Major phases

Process number	Customer deliverable	Definition
D1	Conceptual definition package	This is the collection of the deliverables of the subphases in D1. It contains sufficient detail for a project plan to be developed. It is a document confirming that the specification as defined in D1.4 will meet customer needs and that the product design is in harmony with those requirements and the estimated manufacturing cost.
D2	Launch review package	This is the collection of the deliverables of the subphases in D2. It contains sufficient detail for management to approve a program to be launched. It will include documents defining the cost and schedule aspects of the program and infrastructure requirements that will meet the technical objectives of the program.
D3	Preliminary definition package	This is the collection of the deliverables of the subphases in D3. It contains sufficient detail for management to approve the design concept at the PDR and allows the program to proceed into the detail design phase.
D4	Critical design review package	This is the collection of the deliverables of the subphases in D4. It contains sufficient detail for management to approve the design at the CDR and allows the detail definition data to be released.
D4	Design freeze confirmation	This document authorizes that the product development specs have been approved. Any changes to them will have to follow a rigorous engineering change process. This EC process follows a set of hierarchical authority guidelines.
D5	Product definition package	The product definition package contains all the data required to fully define the physical attributes of the parts and assemblies. These data allow the parts and assemblies to be built by manufacturing.
D5	Product data set	A data set defining all of the part geometry, materials, processes, geometric data (as in plural datum), tolerances, key characteristics, and other necessary dimensions required to manufacture the parts.
D5	Product spec records	A two-dimensional representation of the manufacturing data set plus necessary information for inspection of the finished parts and storage as supporting data for product certification.
D6	Regulatory approval	A statement by a regulatory body that the design complies with all applicable regulations and may be sold and operated in revenue service in that jurisdiction.
D6	Product documentation	All the documents and manuals required for the end customer to operate the aircraft in a safe, efficient and economic fashion.
D7	Program completion review package	This is the collection of the deliverables of the subphases in D7. It contains sufficient detail for the customer to accept the product.
D0	Program plan	This consists of workplan, schedules, integrated master plan/schedule, organizational responsibilities, operating budgets, and a tracking system.

Table 2-4: Customer deliverables/definition example – Subphase

Process number	Customer deliverable	Definition
D1.1	Advanced design study report (aircraft example)	This document contains overall characteristics for the end product including aerodynamic shape, propulsion requirements, interior configuration, performance requirements, and structural arrangements in sufficient detail to show compliance with the market need defined in the marketing requirements and objectives.
D1.1	Marketing requirements and objectives report	This document defines the attributes of a product that meets the customer's economic needs.
D1.1	Benchmark report	This document summarizes available and competitive products and that is used to define the level of technology economic performance and product features needed for a competitive product.
D1.2	Strategic technology reports	The deliverable includes a strategy for manufacturing technology/material technology development on the product [i.e., high-speed machining, composites, carbon brakes, brake by wire, noise systems, smart structure, and computational methods (CFD/CATIA etc.)]. The types of technology required to meet long-term market requirements for cost control.
D1.2	Five-year strategic technology plan	This document defines the opportunities for exploitation of developing technologies within the next five years.
D1.3	Product/process design requirements and objectives	This comprehensive document defines the functional requirements for the end product, not its form. It covers the general design requirements, the product operational and customer requirements, the product design requirements, and the development and qualification requirements.
D1.3	Design standards manual	This document collects the design practices, standardized components, and manufacturing processes to be used for the product.
D1.3	Drafting manual	This governing document defines how the design definition will be communicated to the internal and external customers and how the type data will be stored as indelible record.
D1.4	Product specification document	This comprehensive document defines the functional requirements for the end product, specifically its form. It covers the general design requirements, the product operational and customer requirements, the product design requirements, and the development and qualification requirements and establishes the quantitative quality measures. The structure of the specification reflects the work breakdown structure.
D1.4	Product requirements and capability plan	This document defines the constraints on the product design, due to physical plant capacity and capability. It also describes manufacturing processes not currently available but that could be economically introduced for the program or that can be economically procured from outside sources.

(Continued)

Table 2-4: (Continued)

Process number	Customer deliverable	Definition
D1.4	Regulatory compliance plan	This document describes in detail how the product meets the requirements of an external regulatory body.
D1.4	Technical requirements document	This document defines a product or system requirements. It is used in conjunction with the procurement contract as the basis for procured components.
D2.1	Integrated master schedule	This integrated schedule is cast in the D0–D7 structure, showing major reviews and milestones, with lower level supporting data including major milestones, minimilestones, drawing release curves, production delivery plan, and advance procurement requirements.
D2.1	Human resource plan	This plan is phased with the project schedule and defines the human resources required for project execution and the method for acquisition. It includes projected headcount by organization, critical skill requirements, recruitment plan (internal/external), list of candidates for key positions, training requirements, and plan.
D2.1	Financial plan	This plan is phased with the integrated master schedule and defines the financial implications of program execution in a format useful to both program and functional management. It includes projected cash flow, direct hours per month (total), capital asset budget (for month), unit production cost curves (labor, material), projected sales revenue, and estimate at completion. It also includes capital expenditure requirements for computing equipment, facilities, office space, and infrastructure.
D2.1	Manufacturing plan	This document is based on the product specification and defines the strategy for economic implementation and manufacture. It includes the component and entire product processing requirements, assembly requirements, and tooling.
D2.1	Program partnership plan	This document defines how partners and subcontractors will play a role in the project and the relationship between the partner and the lead organization. The plan includes the partnership philosophy, including the generic terms and conditions of the contracts, the list of candidate partners, and the make/buy policy. For launch review, a set percentage of the material cost and technical requirements must be met.
D2.1	Marketing requirements and objectives product specification matrix	This document shows that the product as specified will meet the market requirements defined in the marketing requirements and objectives report.

Table 2-4: (Continued)

Process number	Customer deliverable	Definition
D2.2	Business philosophy	This document describes the business objectives of the company and relates these to the product development requirements.
D2.2	Product information management structure	The PIM structure identifies the processes and software (PDM) that will be used to manage product information. This covers the life cycle of the program.
D2.2	Work breakdown structure	This document is a breakdown of all work for the project tying in to the overall project WBS. The design tasks should be broken down according to process architecture – family tree that displays and defines the products to be produced in a program and relates the elements of work to each other and to the end product.
D2.2	Preliminary product structure	This document integrates all aspects of the product at the conceptual design stage. This shows compliance with the product specification paragraph by paragraph and how each component is to be manufactured with the cost drivers identified. This may be shown via reports and/or drawings.
D2.2	Computing requirements plan	The capital asset acquisition request, with supporting documentation, defines the computing hardware and software necessary for completion of the program. It must also show allocation of current inventory to the project.
D2.2	Project software standards	This document defines the software standards for the program such that all analyses, interuser communication, and data transfer is via common languages.
D2.2	Facilities plan	This document contains the factory layout plan, the manufacturing process strategy, the shop space requirements, the factory workflow, and the implementation plan. It also contains the office layout/space requirements, key office relationships, and implementation plan all linked to the program schedule. The capital asset acquisition request and supporting documentation are appended.
D2.2	Program-specific business process architecture	This document defines all the program-specific business processes required by the customer or program management. These may include ○ Processes for implementing general design principles ○ Processes for meeting customer requirements for product operations ○ Processes for implementing product design principles ○ A report on applicability of a generic IPD process architecture ○ New processes required by change in the integrated enterprise implementation plan

(Continued)

Table 2-4: (Continued)

Process number	Customer deliverable	Definition
D2.3	Configuration management plan	This document defines the policies and procedures to manage the configuration of the product. It includes policy guidelines, a preliminary drawing tree, or product structure, the database structure, and the program information structure.
D3.1	Systems architecture	This description of a system, or systems, normally in functional block diagram format, depicts top-level system logic, including inter- and intrasystem dependencies. They are presented without regard to physical location and provide less detail than a schematic.
D3.1	Systems layout	This document defines how the various subsystems in the product relate to one another. These may include electrical, hydraulic, wiring, mechanical, etc.
D3.1	Systems performance spec	This document for each subsystem and total system describes the performance expectations. These expectations represent the product in full use by the customer.
D3.1	Vehicle lines and geometry	This definition (from aerospace and automotive) lays out the vehicle mold lines and the geometry of the product. This is completed in conjunction with system layout.
D3.1	Structural interfaces	This document defines the relationship between the key structures within the product (e.g., the relationship between the wing and the fuselage).
D3.1	Internal structural loads	This document is based on analytical study of the key structural members that support the vehicle. Stress analysis is performed to determine strength.
D3.2	Matrix of preliminary definition vs. product spec	This document systematically compares various critical design parameters.
D3.2	Test report	Proof of concept testing (approval at launch review) shows that the conceptual design is valid. It is a test used to confirm that an incumbent technical direction is conceptually sound.
D3.2	DFMA worksheets	This document includes worksheets that are created for each component and assembly to indicate the most efficient design to allow the most cost-effective manufacturing process.
D3.3	Product measures	This document includes four complementary analyses, forming an integral part of the design process, that are primarily used as a measure of design efficiency and cost effectiveness at each design iteration stage and that can be used to indicate a preferred design direction.
D3.3	Trade study documentation	This document describes why decisions were made for historical purposes and to justify design decisions.

Table 2-4: (Continued)

Process number	Customer deliverable	Definition
D3.4	Updated program plan report	This document updates the program plan in light of the refined conceptual design package. Its specific focus is on the recurring cost/nonrecurring cost aspects of the program, to confirm product optimization studies.
D4.2	Matrix of detail definition vs. product specification	This document contains the results of structural and system testing, showing compliance with all requirements. It contains analysis of the physical product vs. product specification, product test reports system test reports and component test reports.
D4.3	Test reports	This engineering report documents the test, the test data, the analysis, and the conclusions.
D4.4	Draft compliance plan	This report lists all paragraphs of the applicable regulations with a statement of who is responsible for showing compliance and how and where proof can be found. It is the first issue of the certification compliance program, circulated for functional impact.
D5.1	Preliminary release package	This package contains all the documents required for releasing drawings to manufacturing and that includes nomenclature, process sheets, assembly manuals, and tool design drawings.
D5.2	Bill of material	This document contains the assignment of numbers to the BOM, which coordinate major components, types of manufacture, and need date for drawing production. It includes the bill of material and product structure.
D5.2	Process sheets	Defined by MRPII, this documents contains definitions of how parts are made, manufacturing process shops, and routings. It also includes detailed process descriptions showing how standard repetitive processes are performed.
D6.2	Certification test report	This is the document that includes test results that form part of the product-specific type data.
D6.2	Vendor qualification reports	This is the vendor's engineering reports that document the proof of qualification testing and compliance testing.
D6.3	Statement of compliance	This document contains a list of all applicable regulations, the means of compliance, a compliance statement, and a reference to the applicable substantiating type data.
D6.4	Maintenance and repair manuals	This document contributes electromagnetic design protection information to the applicable maintenance documentation, including any scheduled maintenance, and electromagnetic design protection information to the applicable repair manual (e.g., repair of lightning/HIRF protection coatings).

Table 2-5: Example of milestones – Aircraft example

Milestones	Definition	Context	Deliverables completed at this milestone
M1 Feasibility study	Feasibility study review is a milestone at which we review the feasibility of the proposed aircraft program.	At this point in the program the conceptual design and competitor comparison studies have been completed to a point where a representative business case can be reviewed.	○ Conceptual design study report ○ Competitor product comparison ○ Aerodynamic lines ○ Feasibility readiness review requirement document
M2 Specification review	Specification review is the milestone at which the aircraft-level specifications are reviewed to determine their capacity to meet the program goals.	In general, this milestone marks the completion of the conceptual design phase. At this point in the program, the technical description of the product is developed in sufficient detail to proceed with preparation of the project plan for the overall project (i.e., schedule, cost, and scope, etc).	○ Product specification ○ Design requirements and objectives ○ Preliminary master lines ○ Conceptual layouts are complete
M3 Authorization to offer	Authorization to offer is the milestone at which the conceptual design, business case, and project schedule are sufficiently established so that the new product can be offered for marketing.	In general, this milestone marks the finalization of the technical description of the aircraft, and the technical risk assessment has determined that a product offering is acceptable.	○ Technical description documents ○ Top-level engineering schedule
M4 Launch review	Launch review is a milestone at which we review the readiness to launch the program.	By this time, the product configuration and its performance are satisfactorily known to a level that ensures that development will be achieved within schedule and cost and that the product will meet customer expectations.	Business case ○ Most technical requirements documents are released ○ Suppliers have been selected In addition: ○ Resource estimates ○ Statements of work ○ Engineering project plan

Table 2-5: (Continued)

Milestones	Definition	Context	Deliverables completed at this milestone
M5 Preliminary design review	Preliminary design review is the milestone where the design is sufficiently complete to evaluate its acceptability to meet the technical requirements and cost, weight, and performance targets.	By this time, the conceptual design has progressed to the stage of detailed layout drawings, and the certification plan has been established.	O Master lines drawing O Detailed layout drawings O Interface control drawings O Certification compliance plan In addition: O The initial board meeting with the regulatory agency has been conducted
M6 Design definition exit review	Design definition exit review milestone marks the end of the joint definition phase. A formal review is initiated to ensure that all preliminary designs are complete and that all suppliers are prepared to return to their facilities with their work share.	By this time, the digital product definition is sufficiently detailed to facilitate the proper integration of all downstream efforts.	O Detail definition layouts and interface control drawings are frozen O The 3D electronic mock-up is established
M7 Critical design review	Critical design review is the milestone that marks the review of the final design. In general, all detail part, assembly, and installation drawings are complete. The finalized design is reviewed against its acceptability to meet the technical requirements, marketing requirements and cost, weight, and performance targets.	By the time, the design has progressed to the stage of detailed drawings, and advanced material orders have been issued. The detail design is frozen following the CDR.	O Detail part O Assembly O Installation drawings O Loads profile O Stress documents

Chapter 3 Table

Table 3-1: Task planning process descriptions

1.0 Process definition	Description	○ IPD task planning is the process wherein programs develop and monitor their program plans. Following authorization, the IPD task planning system is made ready to accept the project information, establishes charge numbers, and launches project-planning efforts. Contractual objectives are loaded into the IPD task planning system, including contract budget and deliverables. Based upon information contained within a program planning package, IPTs within the program develop an integrated implementation plan. The plan is analyzed, reviewed, and approved as the operating baseline for the project. Following approval of the plan, activity status is maintained including cost to complete and incorporation of schedule revisions. Once analyzed, program plans are reported utilizing methods that support the diverse needs of IPD task planning customers. Operating plans are provided at the program and the at the company level.
		○ The IPD task planning process establishes the capability to maintain data to be used by all projects, provides for system and process usability and training, ensures system-level controls and security, and provides procedures for maintenance of the IPD task planning system including modifications and interfaces.
	Objective	○ To provide program and functional management an efficient and effective process and toolset for developing and monitoring program plans that support the delivery and financial objectives of the company's programs.
1.1 Project initialization	Description	○ This process involves establishing projects in the IPD task planning system, establishing plans and authorization for initial program planning and execution activities, and launching the development of the program's operating plan.
	Objective	○ To establish the projects in the IPD task planning system, plan and authorize initial work, and launch development of the programs operating plan within a specified time frame (considering the type of project) after receipt of an authorizing document.
1.1.1 Setup project	Description	○ Assigns specific parameters and guidelines within the IPD task planning system that establish the foundation for maintaining IPD task planning data integrity. This process establishes the project architecture, which is all the projects in the IPD task planning system that collectively make up the program plan.

Table 3-1: (Continued)

		It also establishes some of the project preferences that govern how the project management (Artemis) applications operate. This process involves establishing project-peculiar parameters such as project start and finish dates, planning calendars, resource pools and EAC calculation methods in IPD task planning. The assignment of project IDs and descriptions and the identification of project-specific WBS and IPT structure are examples of project-peculiar information. The unique code fields for all projects in the program are established to ensure the consistent sort, select, and summarization of program planning data.
		○ Some parameters established by the IPD task planning system administrator are required prior to execution of this process. Specifically, the performing department structure, cost element structure, global calendars (cost view), default activity codes, and project codes.
	Objective	○ To create or modify the projects in the IPD task planning system and to establish parameters for enabling the monitoring of a project using the IPD task planning process within a specified item frame (considering the type of project) after receipt of an authorizing document.
1.1.2 Plan, review, and approve initial activity	Description	○ Upon receipt of an authorizing document, project activities that must begin immediately will be planned, reviewed, and approved by program management. Charge numbers for proposal prep or program planning activities and efforts directly related to production of the contracted deliverable product will be established for work that must begin immediately. These work packages will be rolled into the complete integrated implementation plan for projects developing such a plan. Charge numbers will be provided to individuals performing the tasks as well as to systems that accumulate actual changes.
	Objective	○ To identify and plan schedule and resource loads, the initial activities to be performed during the startup of project; to get IPT approval; and to establish initial charge numbers for the task. Process should complete within a specified time frame (considering the type of project) after receipt of an authorizing document.
1.1.3 Obtain, distribute program-planning package	Description	○ Ensure that all the necessary information is accumulated and distributed as reference material from which to develop an integrated implementation plan. The program planning package will include information necessary to identify the following: the projects in the Artemis applications, the valid charge numbers for planning, current program master schedule, financial objectives, and references to all available statements of work and change descriptions. The package also includes the time frame for establishing the implementation baseline plan.
	Objective	○ To provide the necessary programmatic information to support the development of a detailed integrated implementation plan for the program within a specified time frame (considering the type of project) after receipt of an authorizing document.

(Continued)

Table 3-1: (Continued)

1.2 Establish contract baseline	Description	○ Contract baseline is composed of two elements – the contract budget baseline and the contract deliverable baseline. ○ The contract budget baseline is equal to the proposal of record upon authorization. It is adjusted only upon contractual changes. ○ The contract deliverable baseline will be maintained in IPD task planning when detail activity schedules are developed in IPD task planning for a project. Contract deliverables serve as an ending constraint to the project or pieces of the project.
	Objective	○ To establish a baseline for reporting performance to contractual negotiations. ○ To establish planning objectives during the development of an integrated implementation plan. ○ To use contract baselines as source data for program management review of implementation plans.
1.2.1 Load contract budget baseline	Description	○ Loads the time phased budget values, as negotiated in the contract, into the appropriate fields in the IPD task planning system. Contract budget values, in hours and dollars, are identified to the lowest contractual reporting levels of the WBS. These values establish the performance measurement baseline against which performance is reported to the customer when required. Contract budget values are not adjusted except by contractually authorized changes. Identification of the source of changes to the contract baseline is maintained.
	Objective	○ To load the contract budget baseline and validate within two working days of establishment of the project in the IPD task planning system. ○ To establish the contract budget baseline accurately in IPD task planning against which performance will be reported to the external customer. ○ To establish a frame of reference to analyze actual performance to estimated performance to support improved estimates on future contracts.
1.2.2 Load deliverables and program milestones	Description	○ Establish contract deliverable dates and other program milestones as constraints in IPD task planning system as reference point to calculate need dates for meeting those objectives. Provide visibility into deliverables that are in jeopardy of slipping schedule.
	Objective	○ To establish deliverable dates and milestones within two working days of setting up a project in the IPD task planning system.
1.3 Create integration implementation plan	Description	○ Using the information provided within the program planning package, a detailed implementation plan is developed that will become the program operating plan. Creation of implementation plans for the program involve developing integrated plans within each WBS and integrating activities across WBS elements. WBS elements are segregated into work packages and planning packages that provide the total cost to complete for the WBS. Activities within the work planning packages that can be discretely identified are planned as are some activities that may not be specifically identifiable to a work planning package, yet are needed for activity integration or program management visibility. Cost and schedule become integrated through the application of resource values to activities scheduled within work packages.

Table 3-1: (Continued)

1.3 Create integration implementation plan	Description	○ In preparation for establishing the program operating plan, WBS implementation plans are analyzed across the program, and discrepancies are resolved by program team members and management and by functional departments supplying resources.
	Objective	○ To ensure rapid and accurate plan development. ○ To ensure establishment of a plan that meets project objectives. ○ To ensure the integration of cost and schedule. ○ To establish a means to maintain activity integration throughout the project. ○ To identify high-risk areas of the project with respect to resource availability, schedule-critical activities, and cost-critical areas.
1.3.1 Load and integrate tasks	Description	○ This process includes the identification and collection of all work planning packages and activities to develop an entire program plan for the tasks identified in the authorizing document. ○ All direct program resource requirements will be identified during the execution of this process. Labor resources will be identified by the performing department. ○ Information is supplied in all the appropriate code fields at the activity and work package level to support program and process visibility. Some of these fields provide keys for system interfaces. Activities and/or work packages will include span times that reflect the appropriate duration of the task on the project. Activities that roll up to work packages will carry resource values or weights which support the objective measurement of progress. ○ Activity integration is established for activities within and across integrated product teams. ○ Where appropriate, electronic interfaces will be made to receive schedule information from some systems external to the IPD task planning system.
	Objective	○ To identify, load, and ensure integration of all activities and work planning packages to encompass the entire program in sufficient detail to establish an operating baseline. ○ To provide electronic interfaces to receive schedule information from selected systems external to the IPD task planning system, enhancing efficient plan development.
1.3.2 Interpret and analyze plan	Description	○ This process involves the utilization of various methods to analyze and adjust the program plan as it is developed. It establishes analysis methods supporting plan analysis using a multiuser client server environment and establishes a sequence for the most efficient and effective review of a program plan. The process also ensures that all entities needing to analyze the program plan have the methods to support their involvement.
	Objective	○ To identify quickly the key risk areas of the program plan. Areas of risk include potential for missing committed deliverable dates, manpower requirements that are not realistic, span times that are inappropriate for the task to be performed, and missing or incorrect activity integration. ○ To obtain agreement of an integrated implementation plan through a certain level of program management. Includes agreement to assume risk where disconnects are unresolvable at the time.

(Continued)

Table 3-1: (Continued)

1.4 Status, analyze, and Report	Description	◯ This process defines how projects are statused, analyzed, and reported on a recurring basis. This element will define the processes of identifying anticipated or actual start and/or finish dates of activities; accumulating actual costs; identifying revision to the planned activities or work planning packages; adding or deleting work planning packages as necessary; performing analyses considering costs, schedules, and resources; and providing for the reporting of project status and process metrics including various types of summarization. ◯ The measurement and analysis of cost and schedule performance to the operating plan and contract baseline will be supported. Performance to operating plan for all discrete tasks will be measured by an objective method (i.e., milestones assessment and standards). Other tasks will be measured using LOE or apportioned methods as appropriate. ◯ Electronic system interfaces will be established to provide efficient and accurate means for gathering status. ◯ Provides the capability for functional departments to have visibility into resource actual equivalent heads and projected requirements for individual projects and across all projects. ◯ Provides analytical tools and methods to accomplish project performance assessments (including what—if analysis).
	Objective	◯ To provide the processes quickly and accurately to gather project status information, perform analysis, and provide to IPD task planning customers useful status information to support management decisions.
1.4.1 Status implementation plan	Description	◯ Performs the tasks necessary to reflect the current status of all program plans. It involves the collection of actual and expected starts and completions for all activities, work packages and planning packages within the program plan; collection of progress in the form of percentage complete against open work packages; updating percentage complete or remaining duration for detail level activities that are part of network diagrams; updating cost-to-complete at the work planning package level; and calculation of earned value of all open work packages. Closure of work packages upon completion is also performed during this process.
	Objective	◯ To provide program plans that accurately reflect the latest status on a weekly, monthly and quarterly basis as appropriate. ◯ To maximize efficiency by emphasizing automated status methods using interfaces to operating systems.
1.4.2 Accumulate cost	Description	◯ This process collects the actual costs in hours and dollars charged against all accounts in a program. Cost accumulation will be performed electronically through system interfaces to labor accumulation systems (weekly labor) and the cost ledger (total costs). In some cases, actuals may be accumulated by serial number for scheduled activities within work packages as needed for process measurements.
	Objective	◯ To accumulate electronically all costs and apply them to work packages for all accounts within a program. Cost accumulations are to be pulled weekly for labor and monthly for total program costs.

Table 3-1: (Continued)

1.4.3 Incorporate task revision	Description	○ Identify and incorporate revisions, additions, and deletions to the planned activities, work packages, and planning packages where the change is not contractually directed. Also involved is the rolling of planning packages into work packages.
	Objective	○ To incorporate all changes in program plans to reflect the current planning information being used to manage the execution of the program. Ensures accurate activity and work planning package schedules, updates cost-to-complete, and maintains activity integration.
1.4.4 Analyze cost/ schedule	Description	○ Performs the analysis of the program cost, resource, and schedule information identifying and resolving errors, finding risk areas, and performing forecasts based on past performance. ○ Areas of risk include potential for missing committed deliverable dates, manpower requirements that are not realistic, span times that are inappropriate for the task to be performed, and missing or incorrect activity integration, status, or cost accumulation. ○ Analysis should include cost and schedule performance indices, CPM analysis, cost/schedule/performance variance analysis, and estimates at completion forecasts.
	Objective	○ To identify and correct errors quickly, highlighting the key risk areas of the program plan.
1.4.5 Report process metrics	Description	○ Generates and distributes through selected distribution mediums the reports or data files as agreed to with the core company process owners.
	Objective	○ To provide reports necessary to assist core process owners in assessing process efficiency within or across programs.
1.4.6 Report program status	Description	○ Produces and distributes through selected media all program planning information needed by program team members, IPD task planning management, and external customers. Develops a standard set of status reporting packages that provide all the appropriate information at various levels of management. Performs requested ad hoc reporting to address special emphasis items as requested by program management.
	Objective	○ To provide accurate and understandable cost/schedules-related program plan and status information to the IPD task planning customers. ○ To maximize efficiency by using standard program status packages. ○ To maximize effectiveness of status information by providing the appropriate set of performance indicators within any given status package.
1.5 Review and approve project	Description	○ This process involves the collection of the program's implementation plan in formats that support program management review, circulation of the project plan, and receipt of authorization through the various levels of program management. It also includes the establishment of a program operating plan (baseline version) and establishment and maintenance of baseline revision versions and latest approved status versions. ○ This process also governs the authorization of individual work packages as appropriate.

(Continued)

Table 3-1: (Continued)

	Objective	○ To provide useful information to identify quickly inconsistencies or areas of concern within a project plan submitted for approval. To establish an operating plan and to maintain versions.
		○ To provide work package approvals.
1.5.1 Assimilate review package	Description	○ Compiles a set of reports that clearly illustrate the cost, schedule, and resource information needed for a thorough review of the program plan. Ensures the visibility of project objectives and highlights the critical areas supporting those objectives. Prepares to provide recommendations regarding problem areas of the plan if requested.
	Objective	○ To provide a review package that contains necessary reports at the appropriate levels of management to support an effective program management review of the project plans.
1.5.2 Obtain approvals	Description	○ Circulates the implementation plan throughout the various levels of program management, facilitates resolution of problems, and obtains approvals.
		○ Establishes and/or maintains the project baseline (operating plan) and current approved status version and ensures project version control.
	Objective	○ To establish approvals at various levels of a project plan.
1.6 Maintain and administer system	Description	○ This process involves the maintenance of information used by all projects planned within the IPD task planning system. It includes the maintenance of items like cost element and performing department structures, pricing rate schedules, and system security and access. Additionally, the development and implementation of modifications to the IPD task planning system, the IPD task planning system user support, and training will be accommodated within this process. Additional administrative tasks will include the maintenance of project calendars, coding schemes and tables, resource pools, templates, and standard activity libraries. This process will also maintain valid interfaces between the IPD task planning system and functional department systems. To provide for consistent interpretation of IPD data, the establishment and maintenance of reporting formats are also covered within this process.
	Objective	○ To ensure IPD task planning system functionality and user support.
		○ To provide IPD task planning system process/system checks and balances through systems integration and standardization.
1.6.1 Maintain system security and access	Description	○ Provides security and access for users of the IPD task planning system. This will be in either single user or role access at required levels. Access will be requested by MS Mail and will require a program manager or core competency group manager's approval. The IPD task planning system administrator will identify users by role based on input from the program manager or delegated representative. The IPD task planning System Administrator maintains security regarding access to GUI mediums by various internal and external customers.
	Objective	○ To provide security and access for users of the IPD Task Planning System.

Table 3-1: (Continued)

		○ To ensure access and revision control security by user and/or roles as applicable for projects, project versions, data fields, and reports to maintain IPD task planning data integrity.
		○ To maintain security regarding access GUI mediums by various internal and external customers.
		○ To establish a disaster recovery plan to ensure no data loss.
1.6.2 Maintain and modify IPD task-planning system	Description	○ Defines maintenance or modification actions that are required to be performed to ensure that the IPD task planning system will continue to evolve to meet the IPD task planning objectives. These modifications could be in the form of major changes to functionality, emergency fixes of existing functionality, or interfaces to other systems. Maintenance includes changes or additions to calendar, pricing rate structures, cost element structures, performing department structures, activity code library, project code library, resource categories, resource pools, and activity ID categories.
		○ The activity database and template library will be maintained by the IPD task planning system administrator. Criteria for the activities and template creation will be made available and controlled by the administrative function.
	Objective	○ To ensure establishment, maintenance, and consistency of parameters necessary for the operation of program plans in Artemis across all projects.
		○ To maintain master activity database and template libraries to ensure availability for use on all projects.
		○ To ensure all system interfaces and functioning as specified so that data required from other systems is obtained to define and status a project accurately.
1.6.3 Support IPD task-planning users	Description	○ Maintain and monitor IPD task planning help desk to ensure a knowledgable response to user questions. Perform ad hoc reporting to provide data in required format to the IPD task planning users. Support a temporary hot line to provide help following each major implementation.
	Objective	○ To provide help desk support.
		○ To support special ad hoc reporting and processing.
		○ To ensure installed enhancement support.

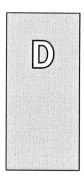

Program Structuring and Planning Checklist

D.1 PROGRAM SCOPE AND ENVIRONMENT CONFIRMATION

○ Confirm program scope and objectives
    ⇨ Review charter, proposal, and existing workplans
○ Confirm cost and schedule constraints
    ⇨ Use planning assumptions worksheet
○ Confirm engineering and work practice guidelines
    ⇨ Use planning assumptions worksheet
○ Interview corporate executives

D.2 PROGRAM STRUCTURING

○ Select and tailor subphases
    ⇨ Use IPD process architecture diagram
○ Select and tailor subphase objectives
    ⇨ Use IPD subphase objectives chart and IPD workflow diagrams
○ Determine customer deliverables
    ⇨ Use subphase objectives chart
○ Select and tailor customer deliverables sections
    ⇨ Use deliverables definition worksheets and IPD workflow diagrams
○ Determine key deliverables
    ⇨ Use IPD workflow diagrams
○ Identify CAD/CAE tools constraints
    ⇨ Use program planning assumptions worksheet
○ Select design techniques and tools
    ⇨ Use program planning assumptions worksheet
○ Select design hardware and software
    ⇨ Use program planning assumptions worksheet
○ Determine design standards and procedures
    ⇨ Use program planning assumptions worksheet

○ Determine replanning points and planning windows
  ➪ Use program planning assumptions worksheet
○ Select program management techniques and tools
  ➪ Use program planning assumptions worksheet
○ Determine program reporting procedures
  ➪ Use program planning assumptions worksheet
○ Determine issue logging and resolution procedures
  ➪ Use program planning assumptions worksheet
○ Perform risk assessment
  ➪ Use risk assessment questionnaire
○ Determine strategies for managing risk
  ➪ Use risk management strategy tables
○ Determine program risk assessment points
  ➪ Use life cycle checkpoints worksheet
○ Tailor the IPD program review process
  ➪ Use IPD *Reference Manual: Program Reviews*
○ Identify program review points
  ➪ Use life cycle checkpoints worksheet
○ Select types of evaluators per review point
  ➪ Use program review selection worksheet
○ Determine involvement of program advisor
○ Determine program decomposition
○ Define roles and responsibilities
○ Create program organization chart
○ Create responsibility matrix
○ Tailor IPD change management procedures
○ Develop integrated master plan

D.3 PROGRAM PLANNING

○ Tailor the workplan templates
  ➪ Review program structuring results
  ➪ Use appropriate workplan template
  ➪ Determine task groups, tasks, and worksteps (select, add, combine, or change tasks and worksteps)
○ Determine task relationships
  ➪ Use workflow diagrams
○ Estimate tasks
  ➪ Use workplan templates (refer to estimating guidelines and estimating units)
○ Adjust task estimates based on program risks

        ⇨ Use risk assessment summary sheet

        ⇨ Use risk factors adjustment estimating worksheet

   ○ Assign resources

        ⇨ Use workplan templates (refer to skill types)

   ○ Develop a schedule

   ○ Tune the detailed workplan

        ⇨ Use program structuring results

   ○ Adjust the integrated master plan

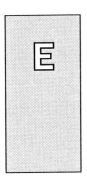

Chapter 8 Tables

Answer each question to the best of your knowledge. You have the option to answer "Not Applicable" to any of the questions, but do so only if the question does not apply to your program, not because you cannot answer it. If you do not have sufficient information, you may overlook a significant risk and run into unpleasant surprises later in the product development effort. Some questions, of course, cannot be answered at the beginning of the program, for example, questions about the use of new CAE technology such as computational fluid dynamics software. In such cases, reassess the risk when the facts are known.

Table 8-1: Risk assessment questionnaire

Example Risk Assessment Questionnaire – Summary Sheet

Program:						

Prepared by:		Date:		Reviewed by:		

Risk Category	Subcategory	Risk Factor	Risk Level L	M	H
Program Size		Manhours			
		Calendar Time			
		Team Size			
		Sites			
		Interfaces to Other Components			
		Organizations to Coordinate			
Program Structure	**Program Definition**	Program			
		Program			
		Benefits of New Product			
		Complexity of Requirements			
		Customer Knowledge			
		Technical Knowledge of Program			
		Availability of Documentation			
		Dependence on Other			
		Dependence of Other Product Interfaces on this			
		Program			
	Sponsorship & Commitment	Commitment of Customer/Partner Management			
		Commitment of Customer/Partner Organizations			
		Relation to Strategic Plan			
	Effect on the Organization	Replacement CAD/CAM New System			
		Effect on Computer Operations			
		Procedural Changes Imposed by the New CAD/CAM system			
		Changes to Organizational Structure			
		Policy Changes			
	Staffing	Program Director Experience			
		Full-time Program Director			
		Full-time IPTs			
		Experience as a Team			
		Team's Experience with Product Technology			
		Team Location			
		Number of Subcontractors			
	Program Management Structure	Methodology Use			
		Change Management Procedures			
		Quality Assurance Procedures			
Program Process Technology	**Hardware & Software**	New or Nonstandard Hardware or System Software			
		Availability of Hardware for Development & Testing			
	Development Approach	New Tools and Techniques			
		New Program Management			
		New DBMS			
	Software Package	Knowledge of Package			
		Prior Work with Vendor			
		Functional Match to System Requirements			
		User Involvement in Package Selection			

Table 8-1: (Continued)

Program size risks	
Risk factor/description	**Risk management strategies**
Manhours – The number of manhours a program involves	
Programs involving a large number of manhours may cause IPT members to lose enthusiasm, become complacent about producing high-quality work, or "burn out."	Perform formal program planning and tracking, supported by an automated program management tool.
	Submit regular status reports to both partner and corporate management.
	Obtain customer acceptance of a well-defined change management procedure at the beginning of the program.
	Use phase development or phased implementation. Partition the program into distinct subprograms, with IPT leader managing each one.
	Reduce program scope.
	Provide for additional program review points.
	Recognize the individual needs of team members.
Calendar Time – The length of the program	
The team's high achievers might become impatient and leave the program, and other personnel might be assigned.	Use the same strategies as listed for the "manhours" risk factor.
Change in the customer business could cause change in requirements.	Use phased development and implementation to show early results and provide key functionality before the complete product is developed.
Change in the customer organization's executive levels could lead to revised priorities.	Although this can be an effective strategy, the overall design concept must still be completed as a unit; otherwise, risks could increase in terms of problems with interfaces between components.
The chance of change in CAD/CAM technology increases with the length of the program.	
Members' sense of urgency is hard to maintain on a long program, potentially leading to a slowdown in pace.	Identify clear milestones and deliverables throughout the program using the IPD process architecture.
The program may become an end in itself for team members, causing them to lose sight of the original business objectives.	Record all design decisions and programs.
	Provide work variety for IPT members.
The team and customer have to wait long to see results.	Use CAD/CAM tools to enhance productivity, thereby reducing the time needed to complete specific tasks. If this strategy means introducing new technology, it may increase risk in that category.
	Overlap activities in the areas of concept design, preliminary and detailed design. Exercise caution in using this strategy because poor timing and management of overlapping activities can lead to other risks.

(Continued)

Table 8-1: (Continued)

Program size risks	
Risk factor/description	**Risk management strategies**
Team size – The size of the program team or IPTs	
If the IPT is too big for one person to directly manage, bottlenecks can occur, causing a slowdown and a slip in schedule. A larger team runs into more communication problems.	Define clear segregation of duties and allocate resource to minimize communication and dependencies between them. Build additional time into estimates and the schedule to provide for communication and coordination. Establish procedures for communicating critical information. Limit the number of IPT members directly supervised to five. This number may be increased to seven, however, if the team is supported by an administrative assistant. Suggest informal procedures to keep information flowing between IPT members. Hold regular staff meetings, with clearly defined agendas and purposes. Partition the IPT into subteams responsible for separate segments of work or deliverables and assign an IPT leader to supervise each sub-IPT directly.
Sites – The number of sites a program has in which the new product will be developed	
When the new product is developed in multiple sites, it may be difficult to define all requirements accurately, particularly if different sites are in different continents, use different hardware or software, or have different policies and procedures. During product development, harmonization difficulties may occur when engineering interfaces.	Establish a PRB with a member from each site and give the PRB responsibility for resolving all disputes or discrepancies among the different sites. This approach necessitates a strong sponsor. Gather requirements from all sites in which the product will be developed. (The requirements may vary across the sites, so an understanding of how each site currently operates is needed.) Additional time must be included in the estimates of the requirements definition subphase to accommodate the extra effort this strategy entails.
Interfaces to other components – The number of interfaces to existing systems	
Improperly defined interfaces from one technology to another (e.g., hardware or DBMS) can result in increased complexity of testing, adverse effects in the systems being interfaced with, and a failure to meet the program schedule.	

Table 8-1: (Continued)

Program structure risks – program definition	
Risk factor/description	**Risk management strategies**

Program scope – Definition of the program scope

Poorly defined program scope can result in a lack of focus by the IPT and in the wasted effort of investigating areas outside the program boundaries.	Obtain customer/partner agreement as to the boundaries of the product development effort. Use workflow diagrams to document the program scope. Review the program scope with the customer/partner and establish procedures for changing the scope after the program begins. Distribute the scope statement and the high-level task relationship diagrams to all members of the steering committee, corporate executives, and IPT. Obtain customer/partner acceptance of a well-defined change management procedure at the beginning of the program.

Program deliverables – Definition of customer/partner deliverables

Failure to define the contents of a customer/ partner deliverable clearly may lead the customer to reject the work as not being what they expected.	Review, with the customer/partner the table of contents of each deliverable in the IPD as well as relevant sample deliverables and reach agreement as to exactly what each deliverable should contain. Review pertinent sample deliverables with the IPT, to guide them in preparing the actual deliverables. Develop a program plan that incorporates the tasks needed to produce the desired deliverables.

Benefits of new product – Definition of the new product benefits

Failure to document and quantify benefits (e.g., in the form of increased revenue, avoided costs, or improved service to customers) can lead to the program being canceled, or future work lost, if the organization faces budget cuts or undergoes a change of management. Lack of clear benefits can also hinder the decision-making process. If benefits are clearly quantified, each product design alternative can be evaluated based on the benefit it provides for the customer business.	Review, with the customer, the features of the new product that will result in cost savings or revenue generation. Document benefits that are of strategic importance to the company, those that will position the organization to establish its business in a new market segment. These benefits are often difficult to quantify and are usually long term in nature. Interview the marketing executives/customers that have the greatest knowledge of the business. Review work papers from prior programs involving the customer in an attempt to identify potential benefits of a new product. Interview the customer support executive, other executives, or program directors who have had prior association with the customer.

(Continued)

Table 8-1: (Continued)

Program structure risks – program definition	
Risk factor/description	**Risk management strategies**

Complexity of requirements	
Customers may not be able to express their needs accurately, thus leading to requirements that are vague and expectations that are difficult or impossible to meet. This leads to a risk to the program. Data needed to design complex products generally originate from a variety of sources, both within the organization and from outside it. If the accuracy of all data is not verified, the resulting product may not meet customer needs.	Utilize prototyping techniques to assist users in discovering the requirements of the new product. Allocate additional time to define and analyze the customer requirements. This will lead to better understanding of the methods of accessing the data to support complex decision-making needs. Have access to people (generally upper and middle management) in the customer organization who will be the product's end users. Provide for additional program review points. Use a database management system with broad query facilities or with an easy-to-use report writer.

Customer knowledge – Customer's knowledge of the product	
If the customer personnel assigned to the program have little or no experience working with the product, a requirements definition may be difficult or yield inaccurate results. This situation is often seen in organizations that have high employee turnover. Decision making and requirements verification can be very slow. It may be difficult to ascertain which features of existing products should be retained.	Insist on access to specialists in the customer's organizations. Utilize senior marketing experts in specific business areas. Allow more time to review documentation. Convince the customer sponsor that a problem in this area exists by documenting the extra time spent and the inaccurate information obtained. Use joint development or a consensus building technique to define requirements. Although the benefits of this strategy can be enormous in terms of joint ownership and consensus building, it can be very risky if an ineffective facilitator runs the sessions or if the wrong people, or too many people, attend. Interview customer personnel who are familiar with any existing products. Experienced customer engineering personnel often have a great deal of knowledge about the products they support.

Table 8-2: Risk assessment questionnaire – Summary sheet

Program structure risks – Program definition	
Risk factor/description	**Risk management strategies**

Technical knowledge of IPT – IPT knowledge about the product technology

The program team's lack of knowledge about the product technology can lead to poorly defined or misunderstood requirements, customer dissatisfaction and frustration, and increases in task completion time and in the time needed for revisions.	Assess the skills of all IPT members, both at the beginning of the program and before the commencement of each program phase. Staff the program team with IPT professionals who have specific product skills or excellent analytical skills. Suggest informal procedures to keep information flowing between IPT members. Have novice and experienced IPT members work side by side on tasks requiring in-depth technical knowledge. Schedule more frequent and formal walkthroughs to facilitate the spread of knowledge throughout the IPT. Allocate extra time in the program plan during preliminary definition, especially for tasks in which team weaknesses will be particularly apparent.

Availability of documentation – Availability and accuracy of documentation of an existing product

Lack of documentation can hinder the progress of the conceptual subphase because engineers cannot quickly gain a good working knowledge of the existing products. Documentation that is poorly written or dated can cause engineers to make assumptions or build new requirements based on incorrect information.	Allocate additional time during the conceptual definition subphase to accommodate the extra effort needed to understand existing products. Find expert users in the customer's organization who have a strong understanding of the day-to-day operation of the existing products. Utilize industry experts who understand the specific technology areas the program involves.

Table 8-3: Risk strategies

Program structure risks – Program definition	
Risk factor/description	**Risk management strategies**
Dependence on other programs – The effect of other programs' progress or deliverables on program	
The lateness of other programs (e.g., another subsystem development effort, the implementation of an integrated product development methodology, a program to establish new standards, the implementation of a database management system, the acquisition of a business, or a relocation of equipment or personnel) could cause the program to miss its deadline. Extra time may be needed to coordinate interprogram issues.	Develop the program plan to account for the program. Identify other IPT's work products or deliverables that are critical to the program. Establish a liaison with other IPTs and schedule regular meetings with them. Share IPT members (across IPTs) for walkthroughs. Review the status reports of other IPTs. Regularly review interprogram dependencies with the customer/partner.
Dependence of other program interfaces on this program – The effect of the program's progress and deliverables on other programs	
The program's progress may not coincide with those of other IPTs. The IPT might suffer if it gives in to pressure from other IPTs to accommodate their programs' needs. In addition, do not be tempted to cut corners by skipping reviews. If the IPT refuses to give in to other IPTs' pressures, the overall customer/partner relationship may be jeopardized.	Use the same strategies (as applicable) as listed for the preceding risk factor. Verify that the work that must be produced for another program is within the scope of the IPT's abilities. Identify clear milestones and deliverables, especially those critical for the progress of other IPTs.

Program structure risks – Sponsorship and commitment	
Risk factor/description	**Risk management strategies**
Program sponsorship – Participation, commitment, and influence of the customer program sponsor	
If the product director is not strong, political battles between business units can result in program delays because of a lack of management commitment. This is particularly true of large programs.	Ask the customer service executive, corporate executives, or quality advisor to intervene with customer management and to communicate the importance of having a strong program sponsor. Ask the customer to elect or appoint a sponsor. Establish a procedure for resolving disputes that occur between different business units of a customer/partner organization. The customer/partner must be responsible for carrying out this procedure.

Table 8-3: (Continued)

Program structure risks – Sponsorship and commitment	
Risk factor/description	**Risk management strategies**

Commitment of the customer/partner management – Degree of customer/partner management's commitment to the program

Customer/partner management's lack of commitment to the program may indicate that they are unaware of the potential benefits of this program, dissatisfied with service, or planning a change in strategic direction.	Review with customer/partner management, all the benefits expected to be achieved by the program. Ask the customer service executive, corporate executives, or quality advisor to intervene with client management and to communicate the importance of the program of the customer/partner organization. Review with the customer/partner management the long-range plan (if one exists), and emphasize this program's role in that plan.

Commitment of the customer/partner user organization – Degree of customer/partner organizations' commitment to the program

Even if customer/partner management is totally committed to the program, users cause serious problems for the development team by refusing to cooperate during the development efforts.	Establish procedures for communicating with representatives of the partner/customer group, including sending them status reports, or scheduling periodic meetings with them. Continue these procedures throughout the design effort, even though minimal partner/customer involvement may be required. Identify preliminary organizational and system impacts and include the partner/customer in that assessment. Communicate the partner/customer organizational concerns to customer/partner management and to the program sponsor. Include customer/partner in walkthroughs during the conceptual definition phases. Use joint application development or a consensus-building technique to define requirements. Since this technique requires the participation of many levels of the organization, it will give the users a greater sense of ownership of the new system.

Relation to strategic business plan – The relationship of the program to an existing strategic business plan

If the program falls outside the scope of an organization's strategic business plan or its implementation deviates from the order prescribed by that plan, risk increases. For example, conflicts within the organization may result because an unplanned program is given priority and resources for it are drawn from other programs.	Review the strategic plan with the customer/partner executive management and ascertain the reasons for developing a product that is outside the plan. If there has been a change in the organization's business strategy, recommend that the strategic plan be revised accordingly. Review other product development programs in the strategic plan to identify any dependencies between the program and ongoing development. Enlist a strong program sponsor.

(Continued)

Table 8-3: (Continued)

Program structure risks – Effect on the organization of a new CAD/CAM system	
Risk factor/description	**Risk management strategies**

Replacement or new CAD/CAM System – Whether a CAD/CAM system is replacing an existing one or is brand new to the organization

If users, operators, and ISD staff are unsure of how the new system will affect their duties, they may resist the work being done (possibly causing a slow down), or they may refuse to use the system after turnover.	Use prototyping techniques to assist users in visualizing the new system.
	Identify preliminary organizational and system impacts and include the users in this assessment.
	Involve representatives of the ISD staff during design, development, and implementation. Ideally, individuals will be selected who will be responsible for maintaining the new system.
	Make sure that the ISD manager is one of the quality evaluators.
	Communicate the new system's organizational benefits (in dollars and cents) to the end users.
	Explain to the end users how the new system will directly benefit them as individuals.
	Develop a detailed plan for involving the users in implementing the new system.
	Develop champions for the new system and keep them involved throughout the development.
	Avoid running both the new and old systems in parallel, if possible.
	Provide training for the new system in phases (e.g. an overview class, a detailed class, and then hands-on demonstrations as needed).
	Use joint application development or a consensus-building technique to define requirements.

Effect on Computer Operations – The effect the new CAD/CAM system will have on the computer operations

In highly integrated system environments, problems may be encountered if the new system must share resources (CPU, output devices, communications lines, databases) or run in a specific time window. Failure to anticipate these issues and to plan accordingly may jeopardize the implementation.	Assess the operations of the data center to determine if it can handle the processing requirements of the new system. Pay particular attention to capacity planning.
	Define new operational requirements, including security, controls, audit, installation, data conversion, back up and recovery, and documentation.
Poor operations documentation and inadequate training for operators may result in difficult and lengthy system implementation. This risk may be even greater if the new system is to run on new hardware.	Require that the operations manager be a member of the management team if the new system's operations will differ drastically from those of the current system.
	Require that the operations evaluator be an experienced member of the engineering operations group and be familiar with the operating procedures of the current system.
	Train the operators and involve them in testing.
	Provide a well-written CAD/CAM operations guide and, if possible, involve the operators in its development.

Table 8-3: (Continued)

Program structure risks – Effect on the organization of a new CAD/CAM system	
Risk factor/description	Risk management strategies
Procedural changes imposed by the new CAD/CAM system – The number of procedural changes necessitated by the new system	
Failure to recognize the need for new or revised procedures and to plan accordingly can greatly increase the risk associated with the program. Users might not take advantage of new functionality because new procedures have not been written to support it. This risk may be even greater if the user has been lax in the development of procedures for the current system (i.e., they may not be in the habit of writing, using, and maintaining procedures).	Review the current procedures and recognize the level of the organization's commitment to procedure development. Prepare an impact statement during the system selection that clearly documents the effects of the new system on existing policies and procedures. Prepare the users as early as possible for changes in their procedures. Begin by reviewing with them the physical data flow diagrams that show procedure changes. Thoroughly test the procedures and associated documentation. Examine the new procedures and determine if any procedures must be changed or written. Review all policy changes and determine if any procedures must be changed or written.

Program structure risks – Effect on the organization	
Risk factor/description	Risk management strategies
Changes to organizational structure – Changes in the new CAD/CAM system could cause changes in the organizational structure	
Failure to recognize and plan for such changes may result in people not knowing their new responsibilities or roles or not being able to hire the right people in time for implementation. Users may not use the new system because they are dissatisfied with the organizational changes.	Prepare an impact statement, during the CAD/CAM requirement definition phase, clearly documenting the effects of the new system on the existing organization. Establish a disciplined approval process for proposed organizational changes, with the program sponsor making the final decisions. Involve representatives from all areas as members of the CAD/CAM committee. Communicate the new CAD/CAM system's organizational benefits to the end users. Develop a human resources strategy, if staff reductions are expected after the implementation of the new CAD/CAM system. Work with functional management in writing new position descriptions.

(Continued)

Table 8-3: (Continued)

Program structure risks – Effect on the organization	
Risk factor/description	**Risk management strategies**

Policy changes – Changes in policy caused by the new CAD/CAM system

A new CAD/CAM system inevitably results in changes in the organization's policies. Because the functional managers make policy decisions, any delay in getting them made could delay the program.	Establish a mechanism to document policy decisions, even minor ones, as they are made.
When large numbers of policy changes are made, there is a risk that some may be lost along the way, ultimately resulting in confusion among users and a system that is less effective than planned.	Assign one person in the organization the duty of coordinating all policy issues (i.e., new, deleted, or changed policies).
	Insist that policy decisions be made by management because they must live with these decisions after the CAD/CAM system is implemented.
	Include an "issues" section in all deliverables, beginning with the requirements report, to focus management's attention on the policy decisions they must make.
	Resolution of these issues can then be tracked throughout the program.
	Recognize that many policy decisions are made during the requirements definition phases.
	Recognize that data analysis will uncover many policy issues that must be resolved before the analysis can be completed.
	Communicate new policies to users.
	Recognize that policies may change as a result of technical changes.
	Review all policy changes and determine if any procedures must be changed or written.

Program structure risks – Staffing	
Risk factor/description	**Risk management strategies**

Product director experience – Level of experience of product director

An inexperienced product director may have difficulties developing an efficient program plan and modifying that plan as the program progresses. This often results in program delays and missed deadlines. On a large complex program, such inexperience can be even more damaging, potentially resulting in increased budget, loss of client confidence, or cancellation of the program or of future work.	Enlist an experienced product director to assist in managing the program.
	Attend program management training.
	Schedule more frequent consultations with the quality advisor.
	Review program plans and status reports from completed programs of similar size and complexity.
	On large programs, use strong program team leaders to complement the experience.

Table 8-3: (Continued)

Program structure risks – Staffing	
Risk factor/description	**Risk management strategies**

Full-time product director – The portion of time that the product director spends on a program

If the product director must manage several subsystem development programs or has duties or responsibilities on other programs, it may be difficult for him or her to focus attention on any of them. This may result in a lack of leadership for the program.	Assign a team member to manage in the absence of the product director.
The customer/partner may lose confidence in the product director.	Contact the team frequently when away (i.e., daily phone calls or voice mail).
It is difficult to solve program-related problems or deal with crises, when not at each working site. Furthermore, there is no buffer between the customer/partner and the team in such situations, and this can lead to inefficiency and frustration among team members.	Let the team know how the product director can be reached in the case of program emergencies.
	Limit the fequency and duration of other responsibilities.

Full-time program team – The portion of the program team that is assigned to the program on a full time basis

When team members split time between several programs, they may find it difficult to focus their attention on any of them. Additional time is required to brief team members on events that have occurred during their absence. The work produced by part-time team members may be more error-prone, thus causing program delays or a poor-quality system.	Insist that the majority of the team spend full time on the program.
The customer/partner may lose confidence in a program team that they perceive as not being fully dedicated to the program.	If team members have duties on another program, limit the duration of those responsibilities. For example, allow them to spend one day a week on the other program for the next two months. Do not agree to allow team members to be available to other programs on an "as needed" basis.
Program team members may become complacent or "burn out" due to the pressures of dual responsibilities or excessive travel.	Develop a mechanism to communicate program details to members who are not on the program full time.
	Assign less critical work to part-time team members.

(Continued)

Table 8-3: (Continued)

Program structure risks – Staffing	
Risk factor/description	**Risk management strategies**

Experience as a team – Amount of experience the team has in working together

When the program team is composed of members who have not previously worked together, delays may plague the initial stages of the program. Team members need time to adjust to other members' personalities; to understand each other's specific skills, strengths, and weaknesses; and to learn how to work together. This risk can be compounded if the team is weak in certain technical skills or lacks the product knowledge needed to complete the program successfully.

On large programs, assign program team leaders who have strong leadership skills.

Attempt to enlist team members who have worked together previously.

Assign responsibilities to team members on the basis of their experience, technical skills, and the type of work they enjoy.

Encourage open communication among team members.

Plan periodic social events for the team.

Allocate extra time during the first several weeks of the program to give the team time to adjust to one another.

Hold regular staff meetings.

Team's experience with the product technology – The amount of experience the product technology team has with the type of product (e.g., jet aircraft, composite wing)

Without such experience, the team will not have the insight necessary to avoid mistakes. Additional time may therefore be required.

Conduct more frequent walkthroughs.

Utilize product specialists as consultants to the program.

Provide training for the team in the particular application area.

Select a team with excellent analytical skills.

Enlist at least one team member with experience in a similar product technology.

Request a senior engineering advisor with experience in that product technology.

Team location–Location of program team members at more than one site

If program team members are working at different sites, the physical distance can cause communication problems. For example, if different sites have different objectives and requirements, additional work can result because of the inability to verify and coordinate the product interface requirements.

Try to move all the team members on a product subsystem (wing) to one location and schedule trips to the different sites as needed.

Schedule regular meetings at a central location and supplement these meetings with conference calls as needed.

Use electronic mail, electronic bulletin board, or networked computer facilities to assist team communication.

Establish procedures to update a central repository of all work produced by team members at different sites.

Insist that all team members, especially those at remote client sites, have work space and facilities adequate to enable them to perform their assigned tasks.

Table 8-3: (Continued)

Program structure risks – Staffing	
Risk factor/description	**Risk management strategies**
Number of engineering subcontractors – The number of subcontractors working on the program	
The team's reputation may be damaged by substandard work done by subcontractors.	Insist on program management control of the subcontractors.
The greater the number of subcontractors on the program, the less control the team has over personnel quality.	Insist that the subcontractors use the same methodologies and techniques as the rest of the team.
If the subcontractor has never before worked on a program with the team, additional time may be required for the subcontractor to learn about the approach to concurrent product development and for the team to learn about the subcontractor's strengths and capabilities.	Investigate the training and experience of the subcontractors.
	Frequently review deliverables produced by the subcontractors.
	Select subcontractors who have worked well with team on other programs.
	Limit the number of subcontractors.
	Establish regular status meetings with subcontractors.
	Insist on formal written contracts with subcontractors.
Methodology use – Use of a foreign methodology or none at all	
Without a standard approach, team members often rely on their own (and generally different) methods of analysis and design development. This can result in poor communication (both with partners and between team members), inadequate estimating and planning, inconsistent documentation, lack of a program review process, and inefficient use of automated tools.	Train the team in the use of the IPD, even if it requires delaying the start of the program.
	When using the IPD for the first time, use it at a high level only. For example, plan at the subphase or task group level rather than at the workstep level.
If a methodology other than the IPD is used, several problems still exist. For example, extra time and effort will be needed for the program team members to become acquainted with the methodology and its techniques and tools.	Tailor the IPD for use on one particular program. Provide automated tool support when possible even though this may increase the technology risk.
However, on programs involving numerous partner personnel who are familiar with their own methodology, it may be less of a risk to learn and use theirs.	When using another methodology, make sure it contains a program review process like the IPD's, one that includes walkthroughs, testing, quality evaluations, and the participation of a program review board.

(Continued)

Table 8-3: (Continued)

Program structure risk – Program management structure	
Risk factor/description	**Risk management strategies**

Change management procedures – Existence of procedures to manage change during the course of the program

Without an effective plan for requesting, tracking, evaluating, approving and implementing changes, you risk losing control of your program. All changes, even minor ones, require additional work by the program team and thus affect the program schedule.	Negotiate a change management procedure with customer/partner management early in the program, ideally during the first two weeks.
	Distribute copies of the change management procedure to all members of the program team and discuss its implications and use with them.
	Manage the changes that occur during the program, i.e., track all changes from request to disposition. (For information on the IPD's recommended change management procedure, refer to the "Change Management," Chapter 11, of this manual).

Quality assurance procedures – The existence of a quality process

If quality assurance procedures are not instituted, chances are great that the system will not meet the customer's expectations and that additional time will be needed to revise defective work.	Focus on customer expectations.
	Do things right the first time. Concentrate first on doing the work correctly; then, focus on doing it faster.
	Provide for training. Training is an essential part of improving quality in concurrent product development.
	Select capable quality evaluators.
	Educate the customer/partner and the program team in the quality assurance process and reinforce these concepts throughout the program.
	Utilize the quality advisor throughout the program.
	Use walkthroughs to find and correct errors early. Incorporate these sessions in the program plan, allocating time for walkthrough preparation by all participants and for correction of all errors discovered.

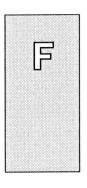

Understanding the ERP and the PDM Connection*

ERP and PDM have different origins and different purposes.

MRP (materials requirements planning) came into being in the 1960s as a methodology or practice for maximizing the efficiency of production through production schedules and inventory control. Even though the definition has evolved with the addition of financial information to MRP II to ERP (enterprise resource planning), the basic functions remain the same – tell me how much inventory and manufacturing resources I will need to produce x amount of product over y period.

MRP software systems evolved to handle the large amounts of data and the complex calculations required to answer this question. They have served as planning tools for production, and accounting tools for reporting inventory status. The late Oliver Wight led the MRP II crusade in the United States beginning in the early 1980s. Practically all ERP gurus today learned the ERP body of knowledge through APICS and "Olie" Wight.

PDM (product data management) came into being in the early 1980s, originally as a means to manage engineering documentation. It offered a number of new concepts just now emerging as mainstream capabilities for general use, such as document management, workflow, and distributed repositories. These "services" were put together to answer questions such as:

○ What does the configuration for the current active revision look like?
○ What did the prior revision of this product look like?
○ What changed between that revision and the current one?
○ How many engineering change orders are outstanding against this product?
○ What is being changed on the bill of material, on the supporting drawings?

The more recent arrival of PDM systems on the scene also means that they are built using more recent technology oriented toward graphical interfaces and distributed user environments.

*From Society of Manufacturing Engineers Blue Book Series, 1998.

2.3.1 The relationship between ERP and PDM

ERP and PDM are integrated at the bill of material (BOM). In fact, since both systems work with BOMs, it is not unusual for there to be some confusion regarding exactly what each system does, especially in regard to the BOM. However, if you look at the basic questions each system tries to answer, it's clear that the BOM is a necessary component for each. To better understand the nature and use of BOMs in both systems a review of how each system uses BOMs is in order.

(i) BOM source

○ BOMs in an ERP system originates in document control or some related function that captures the BOM from engineering and typically processes it through some kind of review and approval process before data entry to the ERP system. When a BOM is entered, it is usually as a completed entity that has been approved for use. Most ERP systems treat this new BOM as ready for production and scheduling based on the effectivity date.

○ BOMs in a PDM system originate out of a CAD tool or systems design as a parts list, either directly or by way of a spreadsheet. The bill is often only partly completed when first entered, and will likely go through several edits before being approved for release to manufacturing.

(ii) BOM usage

○ BOMs in ERP are used to create a production schedule and purchase orders for a set period in time, based on the effectivity dates set in the MRP system for the items in the BOM. As such, they are date oriented.

○ BOMs in PDM are used as a representation of the "as-designed" configuration as it begins to emerge from engineering. PDM serves as a configuration capture mechanism, allowing visibility very early in the product life cycle (ideally well before production release). The configuration goes through various life cycle steps in the PDM system, based on revision control by way of engineering change orders. As such, they are revision code-oriented.

(iii) BOM output

○ The output from an ERP system is a master schedule and materials orders (and perhaps work orders). The BOM is available in report format, by effectivity date.

○ The output of a PDM system is a new BOM for a newly released item or an engineering change order for an existing item. Both are typically driven into an ERP system as a new BOM, or as updates to existing BOMs. The BOM itself is usually presented by revision code.

2.3.2   Users of ERP and PDM systems are different

ERP systems primarily serve accountants, purchasing agents, production planners, shop floor supervisors, shop operators, master scheduler, and all levels of management. The interfaces have been optimized for those classes of users, relying on their specialized domain knowledge to understand how to use the system. In many cases, the primary information vehicle is a printed report or purchase order document. Manufacturing engineers use ERP systems to maintain routing and tooling information. Product engineers use ERP systems to maintain the item master file and BOMs.

PDM systems primarily serve product design engineers, engineering management, document control, change control board members, purchasing, manufacturing, engineering, shop floor users, quality assurance, and field service. As the list of users implies, PDM is focused on distributing product status information to a much broader base of users. It therefore strives for a broader information interface designed to accommodate a wide range of knowledge, skill levels, and uses. The primary information vehicle is on-line graphical BOMs and configuration trees, on-line documents and drawings, and on-line engineering change orders.

> PDM encompasses the management of all forms of digital product data –
> CAD files, geometric models, images, documents, etc. In contrast, MRP
> II systems define and manage textual data.

2.3.3   BOM capabilities are different

ERP is focused on capturing the BOM when the product it represents is ready for production. This becomes somewhat challenging in that lead times for various components may require the product to be scheduled well ahead of when it will actually be built, and that time frame is difficult to predict without first knowing what is in the BOM.

One of the fundamental reasons that organizations bring in PDM is to provide much earlier visibility into engineering change orders and new BOMs, which better facilitates planning and materials management. The underlying driver for doing this is the tendency in most organizations (especially ones with short design cycles and product life cycles) to design right up to the announced production date. More often than not, this results in delivery headaches, quality problems, and manufacturing cost overruns, unless there is configuration visibility almost right at the start of the product life cycle.

PDM is focused on capturing the product configuration as early in the development process as possible. This is to ensure effective use of downstream resources, which depend on timely and accurate product information. The product configuration is the combination of the product structure (e.g., the BOM) and any and all supporting documentation describing and defining the product.

Even though the functionality provided relative to BOMs is extensive in both ERP and PDM, the following list highlights some of the key functional differences:

(i) BOM history

○ MRP has focused on the here and now (show me what I will build in this planning cycle). Most systems do not retain any historical perspective, and those that do have very limited capabilities.

○ The configuration management functions of PDM focus on the entire life cycle of the product; the past (history of a product by revision or date prior to latest released configuration), the present (the latest active configuration), and especially the future (what engineering is planning).

(ii) BOM creation/editing

○ Most ERP systems offer character-based BOM editors, which may be good for high-volume data entry by planners, but are very difficult for others to learn and use.

○ PDM systems, because of the broad audience they serve, have taken a more intuitive graphical approach (although capabilities vary widely between PDM systems) to displaying BOMs for easy interpretation and modification.

(iii) Engineering change management

○ ERP change control is focused on effectivity date designation. Very few offer any kind of revision-based change control – even though that is how the typical engineering change order process is handled. A planner must then convert that information into date-based effectivity for the ERP system. Few if any address the change management process and all the supporting documentation.

○ The focus of PDM is to capture a change request; review and assess the impact in terms of documentation, BOMs, and cost impact; and incorporate that change in the BOM and all supporting documentation for production use.

2.3.4 Limitations of the engineering module in ERP systems

Almost every ERP system has an engineering module, the purpose of which is to act as a staging area for preliminary or unreleased BOMs, and sometimes for management of inventory used in the prototyping phase, and for preliminary costing of BOMs. An engineering change function is almost always a component of this module.

When looking at the challenge of product configuration management (CM) (of which engineering change management is a key part), it becomes easier to see why an ERP "engineering module" does not provide an adequate solution, a notion clearly supported in the excerpt from the APICS professional publication.

> CM, defined simply, is a process that identifies and documents the physical and functional characteristics of an item and manages all changes to the item over its life cycle.

When looking at the broad scope of CM, it must be realized that a PDM or ERP system can manage the total CM cycle by itself.

As a general implementation guideline, consider the PDM system for overall process of engineering change management, interfaced with ERP to provide the information for updating the product structure.

The following list more specifically enumerates the problems associated with using most engineering modules as a basis for engineering change management. Bear in mind that various systems do some of these things better than others, a few do none of them, and no current ERP system does them all. Most engineering modules:

○ Are typically add-ons to appease the engineering need for handling unreleased BOMs without release constraints. They are rarely well designed on the actual requirements for an engineering module in the context of a change manager. Most are also so far down on the ERP suppliers' priority list that they get very little attention or enhancement.

○ Retain the "look and feel" and "assumed expertise" of the core ERP modules, making them inaccessible to most other users in the organization. Any serious attempt to use them beyond the planning group incurs extensive training costs and inefficiencies and mistakes in the business processes supported by the engineering module due to user avoidance. In fact, a common source of errors in the engineering change process, the planning process, and supporting functions such as quality is the difficulty in reading and interpreting complex ERP on-line forms and printed reports.

○ Are typically very costly to deploy to the large group of users who often need access to preliminary configurations, product documentation, and engineering change process.

○ Lack any kind of usable workflow to route the change through an approval process, ensuring that the change to the ERP BOM is the one intended. This causes a serial process requiring extensive human interaction to track and expedite change orders. And it still does not prevent data entry errors.

○ Are not revision driven, which is the BOM view most nonplanning users need to see. This effectively makes them little more than staging areas for pending BOM changes, which still require extensive user interaction to manage and push into the ERP BOM when ready for production.

○ Do not typically handle any of the documents and drawings necessary to communicate and implement an engineering change order (ECO) effectively. This forces the process to continue as a paper-based manual process.

○ Do not provide any historical view of a product configuration. This makes researching change orders very difficult and time consuming, thereby slowing down the process. It also creates significant difficulties for engineering when attempting to reuse a design and for field service when faced with servicing or upgrading older models.

Some engineering modules are not even integrated with the manufacturing module, requiring custom programming or manual procedures to update the production BOM.

PDM defines and manages changes to product data over the life cycle. Change management is process oriented, defining the events in the cycle of reviewing and approving changes. In contrast, ERP systems have focused on the effectivity designation with only limited change management in the workflow context.

2.3.5 Integrating ERP and PDM: Current best practices

A number of organizations have written about ERP/PDM integration. That ERP and PDM are both necessary and should and need to be integrated is not a question any more. Both PDM and MRP II systems are proven techniques in their respective arenas. Each offers capabilities to manage phases of the product data life cycle more effectively. With increasing competitive pressures, it is imperative that manufacturing companies take advantage of all available concepts and information technology.

Amazingly, there are still some companies that manually reenter product structure information from the CAD system in the as-planned version. At a bare minimum, the PDM system should eliminate all manual reentry and use electronic transfer of versions.

APICS offers a number of guidelines for implementing and integrating ERP/PDM solutions, putting an emphasis on tight integration using vendor-supplied tools if possible:

○ From our perspective, usually the most critical interfacing point will be product structure management, including the exchange of item master data. Change management will be a close second.

○ Consider PDM as the server in the client/server environment and, where possible, control all access to product data through the PDM system.

○ Provide automated reconciliation processes to maintain the integrity of the product structure over its life cycle.

○ Provide access without unnecessary log-ins.

○ Continue breaking down the traditional engineering and manufacturing barriers by use of cross-functional development teams and education and team development activities. Develop the concept that data are "our" data, not engineering's or manufacturing's data.

Another even more detailed information source on ERP/PDM integration is the international PDM users group, and independent nonaligned (with either PDM or ERP vendors) organization of PDM/ERP end users. In a June 1996 white paper published by the group, a number of best practices were enumerated. The following is an excerpt from that white paper.

While it is possible to push or pull the "original" BOM and part information from either the PDM or the MRP II system in the interfaced method, it is highly preferred that all information is created in the PDM system and transferred or copied to the MRP II system. This is also true for integration, where the "original" data will live on the PDM system, being accessed regularly by the MRP II system. Intelligent extraction programs can take the needed data from the PDM system and map it to the MRP II systems' input needs. Batch or on-line methods are both easily accomplished.

Part information needs to simultaneously exist in both systems with one updating the other upon changing engineering type data (PDM) through formal EC's (Engineering Change Notices/Orders, etc.), such as revision number, description, commodity class codes, etc. On the other hand, some data entered into the PDM system will never go to the MRP II system, such as much of the supporting documentation, status, approvals and other fields relating to PDM file and record management.

Since product design is started well ahead of production planning/ scheduling, the PDM is the most logical system in which to develop the BOMs. The method for shipping the data to the MRP II system upon a release of some sort should be based upon each organization's procedures.

ECs are also handled far better in PDM systems than in most MRP II systems which gives the PDM system the lead in overall control of the BOM and its parts. The role of the PDM is to push the data to MRP II systems as new product is developed, released, and changed throughout the product life cycle. With this in mind, procedures for release and MRP II database BOM and part updates can be controlled and monitored much more closely than before as a real electronic link contains the data and its updates, ridding ourselves of errors in multiple, untimely, and unauthorized data entry.

Even though the general "best practices" quoted here provide some basis for evaluating and implementing a solution, real-world experience is also important.

How are organizations actually implementing integrated MRP/PDM solutions today?

2.3.6 PDM/ERP integration survey

According to an April 1996 survey of MRP/PDM users by the International PDM Users Group, 71 percent use the PDM system as the master for the BOM. The remaining 29 percent use it as the master for BOM data until the product is released into volume production, and the product enters sustaining mode.

All companies in the survey transfer the BOM electronically, rather than share a common database. The transfer is triggered by a workflow-driven engineering change process. Engineering change orders are also the preferred mechanism for releasing new products into production. For most companies, the transfer is done as a background process on a regular interval basis. Most do not transfer the bill automatically at the conclusion of the ECO workflow but require a final action by a trusted user. PDM systems, probably reflecting their more recent nature, were considered more open than MRP systems, making the integration task more difficult on the MRP side.

Transferred data include both part data and product structure data. Part data attributes always include "description" and "revision." Other attributes transferred by some but not others include

- Unit of measure
- Source code
- Commodity code
- Part type
- Status
- Document ID number w/revision
- Supplier
- Supplier part number
- Notice of pending change on part

Product structure attribute data always include "sequence number" (a.k.a. find number) and "quantity." Other attributes transferred by some but not others include

- Revision label
- Effectivity date
- Reference designator
- Alternate designator
- Variant (option) designation

For effectivity, most companies enter a "target effectivity" date in the PDM system, which can be modified once transferred to MRP. Effectivity is typically tracked against the parent/assembly level. All companies use "in" and "out" effectivity dates.

When BOMs change, the preferred method is to transfer only the changes or deltas, not a whole new version of the BOMs. (The survey noted that most PDM systems do not do this, requiring extensive custom programming to transfer BOM changes to MRP successfully.)

Alternate parts are typically handled by adding a second child to the assembly with the same sequence or find number, but with a quantity of zero. A few did not handle alternates in the PDM system, and one company added alternate parts to a comment field (included supplier and supplier part number).

2.3.7 Tools for integrating PDM and MRP

Up to now, integrating MRP and PDM has been a costly proposition for end users because good tools to do custom integrations were not offered by most PDM suppliers, and there was even less commitment to offer off-the-shelf supported integrations.

A white paper on PDM/MRP integration enumerates several requirements that PDM and MRP suppliers should meet to enable a cost-effective integrated solution. These requirements are excerpted here:

> PDM and MRP II software companies should design their system so that any vendor can deliver data which is organized, identifiable, and usable as input to any other system.
>
> The customer should be able to select what data needs to be transferred/ copied/accessed, and map it accordingly back and forth across a protocol that can manage and monitor it. The customer should not need to write any special code or programs to maneuver through the data except for minimal scripting (e.g., mapping or selecting special fields at the point of transfer, reporting, the use of an interface screen internal to either system, etc.). This could be thought of in terms of templates.

Although the customer is eventually responsible for its own data and applications, it should not be expected to provide major funding nor coding for its applications' data output/input to work with other applications. This does not mean leaning on vendors to produce a product that is enterprisable on the customer's terms as the customer sees the enterprise, not necessarily how the vendor views it.

2.3.8 Conclusion

The competitive agile manufacturer needs control over and access to the configurations for its products throughout their life cycle. Accomplishing this goal requires an integrated ERP/PDM solution, which covers the life of the product from conception through planning and manufacturing. Depending on business practices, a manufacturing execution system may also need to be integrated into the mix. For some companies, this might be extended out to even include field service systems.

Tools and methodologies for integrating ERP and PDM are becoming better defined, driven both by user organizations such as American Production and Inventory Control Society (APICS) and Society of Manufacturing Engineers (SME) and by the ERP/PDM vendors themselves. Enough enterprises have successfully accomplished viable ERP/PDM integration to provide a good body of lessons learned, as well as the confidence that a successful integration now requires little more than good planning.

Manufacturers who achieve reliable and timely control over their product configurations through an integrated ERP/PDM approach will benefit from:

○ Reduced inventory/materials costs, including lower excessive and obsolete inventory, lower materials variances, better utilization of existing inventory, and fewer expediting charges.
○ Reduced manufacturing costs, including lower rework and scrap, more efficient second and third shifts, better factory utilization through faster volume ramps, fewer line stops and switchovers, and faster incorporation of engineering change orders that improve production efficiency.

Increased revenues are initiated through faster production volume ramp-up, fewer lost orders due to slow engineering change cycles for customer-driven changes, and improved customer service.

Glossary

bill of material The list of all materials that go into a component or product. It is organized in a tiered approach starting with the entire product and drilling down to the level of individual items, raw materials, and purchase parts.

body of knowledge Inclusive term that describes the sum of knowledge within a profession or management practice. It includes knowledge of proven as well as innovative advanced practices.

business initiative A project undertaken by the enterprise to improve performance in a given area.

business process framework A structured approach to organizing the business processes of the enterprise. It starts at a high overview level and drills down to the level at which work is performed.

CAD Computer Aided Design.

CAD/CAM Computer Aided Design integrated with Computer Aided Manufacturing. Typically the CAD files will be downloaded to the manufacturing environment such as a numerically controlled (NC) machine or robotic system.

CATIA A brand of three-dimensional product design software that allows virtual assembly of the product being designed.

CDR Critical design review. A formal review of the design following completion of the develop detail definition phase. Customer deliverables are evaluated against the specification and company standards.

CFD Computational fluid dynamics. A mathematically intense branch of fluid mechanics that seeks to simulate the physical performance of a fluid (liquid or gas) in motion.

Change Request CR or ECR. A formal document requesting a change to a product. Change Requests could be used at any point in the product life cycle from conceptual design until the final units are no longer supported. It should include the reason for change, the impact, and the other affected documents.

colocation Establishing a specific physical site where all members of the integrated product team have their work areas during specific phases of the product life cycle. In traditional engineering organizations, the engineering functions are located together under the watchful eye of a lead engineer or supervisor.

collaborative engineering   Projects on which people from different companies come together to form a partnership for the design, manufacture, or support of a particular program. Data and knowledge are shared for the benefit of the design.

concurrent engineering   A design process in which any functional area with an interest in the product have an opportunity to have input into the design up front before the detail product data are developed. The objective is to make the bulk of the product changes up front to eliminate surprises later in the life cycle when they are more expensive to fix.

configuration management   The practice of relating attributes and data to individual parts and then defining the part-to-part relationships for a product. It allows users to find where a component is used across many products and supports version control as an element changes over time.

consensus decision making   An approach used on the Integrated Product Teams where various alternatives are investigated and a solution chosen that optimizes the design across a broad range of variables and inputs from all affected functions.

customer deliverable   The final work product at the highest level that results in a tangible item given to the final customer for approval.

design freeze   Following the preliminary design review, the design is frozen such that no further changes can occur during the development of detail product data.

design to cost   A process whereby the design team places cost targets upon the design early in the process. The team can balance critical product requirements (weight, power, etc.) with cost targets.

DFMA   Design for manufacture and assembly. An engineering approach whereby ease of manufacturing is considered at the point of design. Objectives include reducing part counts and making distinct but similar parts clearly different.

engineering reference manuals   A collection of documents that explains how any given engineering process works. They include such things as drafting requirements, design standards, and approach to project management.

FEM   Finite element modeling. An analytical method for computing scientific results such as stress analysis or aerodynamic analysis that involves breaking a surface down into a grid of elements and analyzing conditions in each element.

functional director   A senior engineering manager responsible for all or a majority of the technical engineering disciplines in an enterprise.

implementation director   Senior manager appointed to direct the overall implementation of the changes in the engineering process. This person will oversee all the subinitiatives and lead the project teams through the transformation.

integrated master plan   IMP. An event-driven plan that describes the activities necessary to design and deliver a product that meets contractual requirements.

integrated master schedule   IMS. Applying resources, cost, and schedule to the detail tasks that make up the integrated master plan.

integrated product development   IPD. A management process that integrates all activities from product concept to field support using a multifunctional team to optimize the product and its manufacturing and sustainment processes to meet cost and performance objectives.

integrated product team   IPT. A cross-functional team used in the design process to allow all affected areas to provide their input to a design up front.

integration team   IT. A team working with the network of integrated product teams responsible for the overall technical, cost, and schedule performance of the program. The IT integrates, allocates, tracks, and manages the inputs and outputs of the IPTs on the program.

key deliverable   A logical grouping of work products that result from actual design activities. Key deliverables are then assembled into customer deliverables, which are then presented to the customer for approval.

life cycle   The natural evolution of a product from concept through design, manufacture, entry into service, field support, and close out.

maturity gates   As a product design is promoted through the design process, it requires that certain elements be in place before moving on to the next phase. Maturity gates control this process.

milestones   Another term for maturity gates, although milestones may not carry the concept of control that a gate provides. Milestones are usually high-level events where top-management decisions are made.

OBS   Organizational breakdown structure. The OBS is the method of organizing the integrated product teams and is usually structured like the work breakdown structure.

organizational change management   The body of knowledge that contains the tools and techniques that allow us to identify and systematically overcome resistance to change. Overcome resistance to change from all levels within an organization.

phase   The first level in the process framework. A phase is a significant piece of work that can be planned, estimated, and managed as a complete project. A phase occurs between two major milestones and is time related.

PDR   Preliminary design review. A step in the design process at which point the conceptual design is formally reviewed against the program constraints and objectives.

PRB   Program review board. Group that provides one of the main links between the customer and the program team. It is responsible for approving the work of the integrated product team throughout the program.

process framework   A structure that defines the logical relationship between business processes, phases, milestones, and deliverables within an enterprise.

process management   An approach that involves thinking of the enterprise as a collection of processes rather than functions. Economic value is created by completing business process. Process management involves studying, organizing, and improving the business processes of an enterprise, often thought of as process reengineering.

process maturity   The concept of measuring process performance and comparing it to standard benchmarks. The standard process maturity model (PMM) consists of five levels of maturity.

product data management   PDM. A tool that helps engineers and the rest of the enterprise manage both the product data and the product development process. It provides a central repository for design data and documentation to ensure that all authorized personnel have access to the most current version.

program   A large package of work resulting in a product ready to be sold. A program is made up of many subprojects.

program director   A senior manager responsible for managing the program and ensuring that it meets technical, quality, cost, and schedule objectives.

program plan   The plan that defines how the program will be completed. It includes the budget, schedule, and scope as well as risk management plans and quality process.

program review   A formal progress review of the program in which outside experts inspect project deliverables and evaluate them using their particular expertise.

project   A package of work on a program that can be planned, executed, and monitored from initiation through wrap up. A project can be the execution of a change request or a minor product enhancement.

project management   The application of knowledge, skills, tools, and techniques to project activities to meet or exceed stakeholder needs.

quality advisor   Participate on the program review board to assess the correctness and completeness of each deliverable from its particular perspective.

quality evaluator   A senior executive who oversees the quality assurance process for the program. It is a proactive role in which the quality advisor counsels the IPT members rather than inspecting their work.

reengineering   A business initiative in which the organization examines in detail how the organization performs tasks to create value. It then optimizes these processes.

ROADMAP Related Objectives and Deliverables Maps. Plan that relates within a project plan how the deliverables relate to create a final Customer Deliverable.

routings The steps within the manufacturing process that a component will follow as it moves through the manufacturing process. In the fabrication process, a part might follow a routing that includes cut, drill, stamp, deburr, and heat treat.

scope The statement of work for a project defining the work that is expected to be completed on the project.

specification The document that defines what the product is intended to be. It outlines key parameters of the design such as size, speed, and stresses.

specification tree The structure of the specification into logical groupings and organized by a common numbered tracking system.

sociopolitical management The practice of managing organizational politics by understanding the role of the individual within the dynamics of the whole organization. This includes integrating social science and political science to achieve change.

statement of work A detailed outline of what work is to be completed on a program or project. Includes the workplan, deliverables, timing, cost and resources.

subphase The second level of the process framework. It defines a logical block of work that is smaller and more manageable than a phase.

subphase workplan The workplan for a subphase that outlines at a detailed level all the steps that will need to be completed to complete the subphase. The steps will be known in such a way that an estimated cost and schedule can be accurately prepared.

systems engineering An organized approach to solving the requirements of designing a complex system. It is an interdisciplinary approach and means to enable the realization of successful systems.

task Subphases are further divided into task groups and tasks. A task is a precise piece of work with a detailed description, design tips, and examples.

task group Subphases are further divided into task groups and tasks. A task group is simply a logical grouping of tasks.

team agreement A written commitment established between the program director and the integrated product teams to provide the necessary framework for team performance.

time to market The performance measure that tracks the time elapsed between the identification of a market need and the delivery of the final product to the first customer.

VPD Virtual Product Development. A process of designing and prototyping a product with a CAD system. Digital models of individual components are assembled on-line, and potential interferences and other design flaws are addressed.

walkthrough A quality process taking the form of a short meeting in which several integrated product team members examine work products with a common goal: to find errors.

work breakdown structure WBS. A logical relationship between all the tasks that need to be completed within the design process. The WBS is usually organized around the product structure.

workflow A documented standard process that controls the way data and deliverables will move through the business process. It might be a standard approval process that defines which members of the team need to sign off on a drawing and in which order this must occur.

workplan The result of completing the work planning template showing the resource requirements, deliverables, and detailed cost and schedule.

workplan template A standard approach for managing the completion of tasks. Workplan templates include for each task: a description, a required skill, the number of units of that skill, estimating guidelines, a skill type, and an estimate of the hours resulting.

work product Deliverable resulting from the completion of a task

workstep The most specific unit of design work. It produces a single work product and can typically be assigned to an individual. An example might be to calculate the moment of inertia.

Bibliography

Allen, L. A. *The Management Profession.* New York: McGraw-Hill. 1964.

Armstrong, S. *de Havilland Aircraft BES Conceptual Design Manual.* AMGI Management Group, Inc., under contract to de Havilland, Inc., Toronto. 1996.

Bedeian, A. G., and Glueck, W. F. *Management,* 3rd ed. New York: Dryden Press. 1983.

Bombardier Aerospace Engineering System (BES). Applied to Regional Jet Program. Toronto: Bombardier Aerospace. 1997.

British Aerospace Integrated Product Development (IPD) System. Developed under the Operation Efficiency Improvement (OEI) initiative. Warton, Lancashire, UK: British Aerospace. 1996.

Charney, C. *Time to Market – Reducing Product Lead Time.* Dearborn, MI: Society of Manufacturing Engineers. 1991.

Davenport, T. M. *Process Innovation: Reengineering Work Through Information Technology.* Boston: Harvard Business Press. 1993.

De Board, R. *The Psychoanalysis of Organizations.* London: Tavistock Publications. 1978.

Dickerson, C. *PDM Product Data Management: An Overview.* SME Blue Book Series. Dearborn, MI: Society of Manufacturing Engineers. 1998.

Drucker, P. *Concept of the Corporation.* New York: John Day. 1946.

   The Practice of Management. London: Pan Books in association with William Heinemann. 1955.

Duncan, W. *Just in Time – In American Manufacturing.* Dearborn, MI: Society of Manufacturing Engineers. 1988.

Ernst and Whinney. *Systems Development Methodology (SDM). Version 3.0.* Cleveland, OH: Ernst and Whinney. February 1989.

Gregory, C. *The Management of Intelligence, Scientific Problem Solving and Creativity.* New York: McGraw-Hill. 1967.

Grint, K., and Willcocks, L. *Business Process Re-engineering in Theory and Practice: Business Paradise Regained.* Oxford, UK: Templeton College, Oxford Institute of Management. 1995.

Gunn. T. *Manufacturing for Competitive Advantage.* Cambridge, MA: Ballinger Publishing. 1987.

Hammer, M. Reengineering work: Don't automate, obliterate. *Harvard Business Review,* July 1990.

Hammer, M., and Champy, J. *Re-Engineering the Corporation.* London: Nicholas Brearley. Harper, Business. 1993.

Juran, J. M. *Juran on Leadership for Quality – An Executive Handbook.* New York: Macmillan. 1989.

Kanter, R. M. *The Change Masters: Innovation and Entrepreneurship in the American Corporation.* 1985.

Kircher, P., and Mason, R. O. *Introduction to Enterprise – A Systems Approach.* New York: Kaelville Publishing Company. 1975.

Lee, G. L., and Smith, C. *Engineers and Management – International Comparisons.* London: Routledge Publishing. 1992.

Lockheed Martin Engineering Work Package and Activity Definition Approach. Integrated Cost and Schedule (ICAS) initiative. Fort Worth, TX: Lockheed Martin. October 1996.

Lockheed Martin IPD Maturity Self Evaluation Tool. Fort Worth, TX: Lockheed Martin, Product Engineering Department. February 24, 1997.

Lockheed Martin Product Development Roles and Responsibilities – Functional Management vs. Program Management. Presentation by Lockheed Martin Product Engineering, Fort Worth, TX. 1997. Virtual Product development initiative.

Lockheed Martin Tactical Aircraft Systems Process Maturity Model. Fort Worth, TX: Lockheed Martin, Product Engineering Department. 1997.

Machiavelli, N. *The Prince,* 1513. Translated by W. K. Marriott. New York: Institute for Learning. Columbia University. 1995.

Management Consulting – A Guide to the Profession. Geneva: International Labour Office. 1986.

Maslow, A. H. A theory of human motivation. *Psychological Review,* 370–96. 1954.

McClinton, D. F. *The Unwritten Rules of Systems Engineering.* 1996. Presented at the Fourth International Symposium of the National Council on System Engineering (NCOSE). 1994.

Metzer, P., and Boddie, J. *Managing a Programming Project, Processes, and People,* Introduction and Chapter 1. Upper Saddle River, NJ: Prentice Hall. 1993.

MRP and PDM: Understanding the Fit – How to Maximize Results with an Integrated PDM/MRP Solution White Paper, Agile Software Corporation. *www.agilesoft.com.*

Naisbitt, J., and Auburdene, P. *Re-inventing the Corporation – Transforming Your Job and Company for the New Information Society.* New York: Warner Books. 1985.

Nicholas, J. M. Concurrent engineering: Overcoming obstacles to teamwork. *Production and Inventory Management Journal,* 3rd quarter, 1994.

Pfeffer. J. *The Human Equation.* Boston: Harvard Business School Press. 1998.

Porter, M. *Competitive Strategy: Techniques for Analyzing Industries and Competitors.* New York: Free Press. 1998.

Pritchett, P., and Price, R. *High Velocity Cultural Changes Handbook for Managers,* pp. 2–10. Dallas, TX: Pritchett Publishing Company. 1993.

Project Management Institute. *Guide to the Project Management Body of Knowledge.* Upper Darby, PA: Project Management Institute. 1984.

Quinn, J. B. *Intelligent Enterprise – A New Paradigm for a New Era.* New York: Maxwell Macmillan. 1992.

Scott-Morgan, P., and Little, A. D. *The Unwritten Rules of the Game.* New York: McGraw-Hill. 1994.

Senge, P. *The Fifth Discipline – The Art and Practice of the Learning Organization.* New York: Currency Doubleday. 1995.

Stafford Beer. *Management Science – The Business Use of Operations Research.* New York: Doubleday Science Series. 1968.

U.S. Department of Defense. *Guide to Integrated Product and Process Development.*

Index

Printed in the United Kingdom
by Lightning Source UK Ltd.
107467UKS00002B/271-290

9 780521 017749